CONSUMER SERVICES AND ECONOMIC DEVELOPMENT

The rapid expansion of the service sector is a principal feature of contemporary global economic restructuring and has prompted a reconsideration of the role of services in economic development. For the most part, however, this rethinking has focused upon one minor aspect of the service sector, namely producer services (services which supply other businesses). In contrast, this book examines the importance of the far larger consumer services sector (services which supply final consumers).

Consumer Services and Economic Development evaluates the role of consumer services as motors for local economic development in contemporary advanced economies. The contributions that specific consumer service industries are making to local economic development are analysed. Tourism, sports, universities, retailing and the cultural industries are each examined in turn. Following this, the roles that the consumer services sector are playing in economic regeneration are investigated in a number of different contexts: a global city; several contrasting urban areas; and a rural locality. In each case, whether consumer services have taken a lead or supportive role in local economic development is surveyed, and their effectiveness in promoting economic regeneration is evaluated.

This book dispels the common myth that consumer services are residual activities dependent upon other economic sectors for their vitality and viability. Providing a comprehensive overview and evaluation of the contributions of the consumer services sector to economic development, this book reveals the need for a fundamental reconceptualization of both the function of services in particular and economic development theory and practice more generally.

Colin C. Williams is a Senior Lecturer in Urban Development and Planning, Leeds Metropolitan University.

CONSUMER SERVICES AND ECONOMIC DEVELOPMENT

Colin C. Williams

London and New York

First published 1997
by Routledge
2 Park Square, Milton Park, Abingdon, Oxon, OX14 4RN

Simultaneously published in the USA and Canada
by Routledge
270 Madison Ave, New York NY 10016

Transferred to Digital Printing 2009

© 1997 Colin C. Williams

Typeset in Garamond by J&L Composition Ltd, Filey, North Yorkshire

British Library Cataloguing in Publication Data
A catalogue record for this book is available from the British Library

Library of Congress Cataloguing in Publication Data
Williams, Colin C.
Consumer services and economic development / Colin C. Williams.
p. cm.
Includes bibliographical references and index.
1. Customer services—Great Britain. 2. Great Britain—Economic
conditions—1993—Regional disparities. 3. Great Britain—
Economic policy—1945- I. Title.
HF5415.5.W582 1997
338.941—dc20 96–41516

ISBN 0–415–14504–X
0–415–14505–8 (pbk)

Publisher's Note
The publisher has gone to great lengths to ensure the quality
of this reprint but points out that some imperfections
in the original may be apparent.

to Jan

CONTENTS

CONTENTS

TABLES

TABLES

ACKNOWLEDGEMENTS

Books, despite having only the author's name on the title page, are more often than not the result of much collaboration by a wide range of people. This one is no different. Without the time and effort of a large number of people, the learning and writing process involved in this endeavour would have been much slower.

In this regard, I should like to thank particularly a number of colleagues who freely gave their time both to read and make incisive comments on draft versions of chapters, namely Bill Bramwell, Ron Griffiths, Paul Lawless, John Montgomery, Michael Pacione and Brian Robson. For helping to collate the literature so as to make the task of writing the book so much easier, I am indebted to Aidan While, whilst Trevor Hart put up with my continual pestering for help with organizing the employment data. Above all, however, I should like to thank Jan Windebank for reading the whole of the book, providing inspiration and making detailed editorial comments. Of course, the normal disclaimers apply: any omissions or faults are mine alone.

The author and publishers also gratefully acknowledge permission to publish the following material:

Addison Wesley Longman Ltd for permission to reproduce material from N. Wrigley and M. Lowe (eds) (1996) *Retailing, Consumption and Capital: Towards the New Retail Geography*; J.N. Marshall and P.A. Wood (1995) *Services and Space: Key Aspects of Urban and Regional Development*.

Avebury Publishing Ltd for permission to reproduce material from G. Haughton and C.C. Williams (1996) *Corporate City?: Partnership, Participation and Partition in Urban Development in Leeds*.

ACKNOWLEDGEMENTS

Blackwell Publishers for permission to reproduce material from P.W. Daniels (1993) *Service Industries in the World Economy*.

Brewers and Licensed Retailers Association for permission to reproduce material from *A Real Alternative: Memorandum to HM Treasury* (1994).

Carfax Publishing Company for permission to reproduce material from G. Dabinett (1991) 'Local policies towards industrial change: the case of Sheffield's Lower Don Valley', *Planning Practice and Research*, 6, 1: 13–18; H. Lim (1993) 'Cultural strategies for revitalising the city: a review and evaluation', *Regional Studies* 27, 6: 589–95; R. Paddison (1993) 'City marketing, image reconstruction and urban regeneration', *Urban Studies* 30, 2: 339–50.

Frank Cass & Co. Ltd for permission to reproduce the table 'Structure of employment' from T. Elfring (1989) 'The main features and underlying causes of the shift to services', *Service Industries Journal* 9: 337–46.

Committee of Vice-Chancellors and Principals of the Universities of the United Kingdom (CVCP) for permission to reproduce Table 7.1 from J. Goddard *et al.* (1994) *Universities and Communities*.

Elsevier Science Ltd for permission to reproduce material from F. Moulaert and F. Todtling (1995a) 'Preface', *Progress in Planning* 43, 2–3: 101–6; F. Moulaert and F. Todtling (1995b) 'Conclusions and prospects', *Progress in Planning* 43, 2–3: 261–74.

Geografiska Annaler for permission to reproduce material from M. Lundmark (1995) 'Computer services in Sweden: markets, labour qualifications and patterns of location', *Geografiska Annaler* 77B, 2: 125–39.

Growth and Change: A Journal of Urban and Regional Policy for permission to reproduce material from J. Harrington, A. MacPherson and J. Lombard (1991) 'Interregional trade in producer services: review and synthesis', *Growth and Change* 22, 4: 75–94.

HMSO for permission to reproduce material from R. Kennedy (1991) *London: World City Moving into the 21st Century*.

Manchester University Press for permission to reproduce material from F. Bianchini and M. Parkinson (1993a) *Cultural Policy and Urban Regeneration: The West European Experience*.

ACKNOWLEDGEMENTS

Pion Limited, London, for permission to reproduce material from J.N. Marshall and P.A. Wood (1992) 'The role of services in urban and regional development: recent debates and new directions', *Environment and Planning A*, 24: 1255–70; C.C. Williams (1996a) 'Understanding the role of consumer services in local economic development: some evidence form the Fens', *Environment and Planning A*, 28, 3: 555–71.

Review of Urban and Regional Development Studies for permission to reproduce material from J.C. Stabler and E.C. Howe (1993) 'Services, trade and regional structural change in Canada, 1974–84', *Review of Urban and Regional Development Studies*, 5: 29–50.

Routledge for permission to reproduce material from D. Gibbs (1989) *Government Policy and Industrial Change*.

Every effort has been made to trace and contact known copyright holders before publication. If any copyright holders have any queries they are invited to contact the publishers in the first instance.

1

INTRODUCTION

The rapid expansion of the service sector is a principal feature of contemporary global economic restructuring. The service sector's share of the world's workforce rose from 24 per cent to 35 per cent between 1965 and 1990, and its share of global Gross Domestic Product (GDP) increased from 50.6 per cent to 62.4 per cent between 1960 and 1990 (International Labour Organization 1995). Recognizing this, economic development practitioners and academics alike have embarked on a reconsideration of the role of the service sector in economic development. For the most part, this rethinking has focused upon the contributions made by producer services (i.e. services which supply other businesses). In stark contrast, few have sought to explore the role of consumer services (i.e. services which supply final demand), even though such services constitute the vast bulk of the service sector. Instead, consumer services are predominantly assumed to be residual activities simply dependent upon other economic sectors for their vitality and viability. Being a by-product of the wealth created in these other sectors, consumer services are thus deemed uninteresting, unrewarding and unworthy of investigation. Indeed, so dominant is this perception that apparently no book has been written before on consumer services and economic development. Contemporary wisdom about this sector, which in advanced economies such as Britain employs some 40 per cent of all employees, instead remains grounded in little more than inference and supposition.

The aim of this book, therefore, is to fill a major gap in our understanding of contemporary economies. Its objective is to evaluate critically the role of consumer services in economic development. In so doing, it will reveal that the study of consumer services, far from being uninteresting and unrewarding, enables a major contribution to knowledge to be made. By examining the evidence on the role of consumer services in economic development without the presupposition that they are parasitic activities, this book not only uncovers the positive contributions of consumer services but also radically reconceptualizes how we think about economic development and how we seek to achieve it in practice. Before going any further, however, this introductory chapter must first define what is meant by services in general and consumer services in particular.

DEFINING CONSUMER SERVICES

At present, the dominant method for classifying economic activities is to divide them into primary (extractive), secondary (manufacturing) and tertiary (service) industries. The problem immediately confronting anyone who wishes to examine services, however, is that despite the fact that this sector now employs 67 per cent of all employees in advanced economies (International Labour Organization 1995), it was originally designed as a 'residual' category to capture all activities which do not neatly fit into primary and secondary activities.

Given this history, it is perhaps inevitable that any attempts to define services by what they are will be unsuccessful. For example, two criteria frequently employed to define a service positively are that the item can only be consumed at the point of production (for example lecture, haircut, restaurant meal) and that the item takes a non-material form (for example consultation with a doctor, live theatre, a seminar). The problem, as Urry (1995) notes, is that some services are not consumed at the point of production (for example take-away meals) and others take a material form (for example software). Today, therefore, the study of services remains plagued by the problem identified by Stigler (1956) over four decades ago: there exists no authoritative consensus of either the boundaries or the classification of the service industries. Although many have strived to identify specific features which characterize all services (Daniels 1985, Marshall and Wood 1995, Riddle 1986), they all recognize that the diversity of both the activities and businesses encompassed under the service-sector umbrella makes an all-embracing definition inappropriate. Services, after all, are an historical legacy of a classificatory schema which attached importance to primary and manufacturing-sector activities.[1] Consequently, such activities, as Marshall and Wood (1992: 1255) assert, can only be defined by what they are not: they are 'activities relatively detached from material production'. Since this is essentially what they are, they must remain defined in this residual manner.

Confronted with this seemingly intractable problem of defining the sector as a whole, many researchers have tended to retreat quickly to the relatively safer terrain of dividing these heterogeneous activities into a series of comprehensive and more internally consistent sub-categories. To do this, the most common approaches adopted have been to classify services according to their markets, tradability across international borders, occupation, or various combinations of all three (for an in-depth discussion, see Daniels 1993, Marshall and Wood 1995).[2]

The most widely used and familiar of these categorizations is the approach which classifies service industries according to their markets. This divides them into four groups: 'producer services', which meet the intermediate demand of other firms; 'consumer services', which fulfil final demand; 'mixed producer/consumer services', covering those activities which satisfy a relatively mixed proportion of both final and intermediate demand; and 'public' services, which meet the demand for government services. In the United Kingdom, the dominant way of allocating service industries to these four groupings is the approach adopted by the Producer Services Working Party during the 1980s (Marshall et al. 1988). Outlined in Table 1.1, this will be the method employed in this book for identifying the scope of the consumer services sector.[3]

Some have argued that the consumer/producer services dichotomy breaks down in practice and is thus irrelevant. Consequently, it should be discarded (for example Allen 1988). This book partially agrees with this. It is correct that service activities do not neatly fit into these categories. Most services tend to fulfil both intermediate and final demand, and the proportion of production destined for each demand category tends to vary according to such factors as firm size, ownership, corporate organization and location. However, this book does not agree that the producer/consumer services dichotomy is therefore completely irrelevant. As will now be shown, this classificatory scheme is so totally ingrained in popular discourse that for further progress to be made in the study of services in particular and economic development more

Table 1.1 1980 UK standard industrial classification (SIC) of consumer services, producer services and mixed producer/consumer services

1980 SIC index	Description
Consumer services	
64–5	Retail distribution
661	Restaurants, snack bars, cafés
662	Public houses and bars
663	Night clubs and licensed clubs
665	Hotel trade
667	Other tourist/short-stay accommodation
672	Footwear/leather repairs
673	Repair of other consumer goods
721–2	Road passenger transport/urban railways
846	Hiring out consumer goods
931	Higher education
932	School education
936	Driving and flying schools
95	Medical/health/veterinary
96(excluding 963)	Welfare/religious/community services
97	Recreational/cultural
982–9	Personal services
Producer services	
61–3	Wholesale distribution/scrap dealing, etc.
723	Road haulage
831–2	Auxiliary services to banking/insurance
837–8	Professional/technical services; advertising
839	Business services
841–3, 849	Hiring out machinery, equipment, etc.
94	Research and development
963	Trade unions, business and professional associations
Mixed producer/consumer services	
664	Canteens
671	Repair/servicing of motor vehicles
71	Railways
726	Other transport
74	Sea transport
75	Air transport
76	Support services to transport
77	Miscellaneous transport services
7901	Postal services
7902	Telecommunications
814	Banking

Continued

Table 1.1 (continued)

1980 SIC index	Description
Mixed producer/consumer services *(continued)*	
815	Other financial institutions
82	Insurance
834	House and estate agents
835	Legal services
836	Accountants, auditors, tax experts
848	Hiring out transport equipment
85	Owning and dealing in real estate
933	Educational and vocational training
981	Laundries, dry cleaners, etc
Public services	
91	Public administration, national defence

Source: Marshall *et al.* (1988: Table 2.4)

generally, an in-depth critical evaluation of the dichotomy must first be undertaken. Indeed, it is for precisely this reason that it is retained here.

RATIONALE FOR STUDYING CONSUMER SERVICES

Many of the assumptions underlying the role of consumer services in economic development are grounded in 'economic base theory' (Haggett *et al.* 1977, Glickman 1977, Wilson 1974). Founded on the idea that an area needs to earn external income in order to grow, this theory divides any economy into two sectors: 'basic'-sector activities, which generate external income for the area and act as the 'engines of growth'; and 'dependent'-sector activities, which merely circulate income within the economy and are perceived as activities which are at best reliant on the basic sector and at worst, parasitic. Given this view of the economy, much of the attention in local economic development is focused upon cultivating the basic sector. Traditionally, this was seen to be composed of primary and secondary activities, whilst the service-sector was relegated to the dependent sector. The major achievement of the vast array of service sector studies conducted over the past decade or so, however, has been to redefine producer services as basic-sector activities and thus motors of local economic development. Consequently, the old 'manufacturing as engine of growth' versus

5

'services as dependent activity' dualism has been replaced by a 'manufacturing and producer services as engines of growth' versus 'consumer services as dependent activity' dichotomy. Remaining largely intact, however, are first, the conceptualization of the economy as consisting of both a basic and a dependent sector and second, the idea that an economy can be structured according to a 'hierarchy of sectors' prioritized by their ability to export. By focusing upon the role of consumer services in economic development, the key purpose of this book is to evaluate critically both these conceptualizations.

First, by revealing that consumer services, as the major sector remaining entrenched in the dependent or residual category, fulfil an external income-generating function, this book raises serious doubts about the validity of continuing to use the basic-/dependent-sector dualism in economic development theory and practice. Hitherto, the problem has been that specific consumer services have been studied in isolation (for example tourism, universities, sport, cultural industries, medical services). This has meant that although individual consumer service industries have been shown by commentators to be functioning as basic-sector activities (for example Bale 1993, Burgan and Mules 1992, Goddard *et al.* 1994, Law 1993, Myerscough 1988, Wynne 1992c), without a re-evaluation of consumer services as a whole, the dualism itself has gone unchallenged. This book, however, by studying the consumer services sector in aggregate, will bring the basic-/dependent-sector dualism starkly into question. Revealing that the consumer services sector as a whole functions as a basic industry, it will contest the very notion of a basic-/dependent-sector dualism.

By investigating the role of consumer services in economic development, however, it is not only the basic-/dependent-sector dualism which will be questioned. The second assumption that remains at the heart of much economic development theory and practice is that sectors should be prioritized according to their ability to export, which results in a 'hierarchy of sectors'. Through the study of the contributions of consumer services to economic development, however, serious misgivings are advanced about this hierarchical approach. Commencing in chapter 4, and with further illustration and explanation throughout the sub-sector and locality studies in Parts II and III, it will be shown that to grasp the full relevance of consumer services in local economies, one has to progress beyond an appraisal of their basic-sector characteristics.

To focus solely upon these characteristics would be to assume that economic growth is merely a product of the level of external income-generating activity in an area. However, research has revealed that economic growth is not so strongly correlated with income-inducing activity as many previously assumed (Giaratani and McNelis 1980, McNulty 1977, Mandelbaum and Chicoine 1986). This is because for an economy to grow, it is not a rise in external income *per se* which is needed but, rather, an increase in *net income*. Net income, to explain, is determined by total external income, times a multiplier (which is higher the more self-reliant the economy), minus total external spending. The growth of a local economy is thus dependent not only on attracting external income but also on preventing the leakage of money out of the area.

Accordingly, locally oriented activities can play an important additional role in encouraging economic growth by thwarting seepage of spending out of the local economy. This can be accomplished in two ways. First, locally oriented activity can prevent money from draining out of the locality by supplying facilities which negate the need for the local population to travel outside the area to obtain a good or service. Second, locally-oriented activity can change the expenditure patterns of local businesses by raising the share of their spending in the locality.

To understand the full contribution of consumer services to local economic development, it is thus useful to envisage consumer services as performing two functions in local economies. First, they can function as basic activities, drawing income into the economy from outside, and second, they can operate as leakage preventers, serving the local market and thus retaining and circulating money within the local economy. These two features can be seen as equally important in the economic development of a locality as both are required to increase net income.

As will be revealed throughout this book, the difficulty at present is that the positive role that locally oriented activities play in economic growth as leakage preventers is frequently under-emphasized in economic development theory and practice. The outcome is that many local economies leak like a sieve. Consequently, rather than concentrate upon those basic activities which generate income for local economies, as in the export-oriented hierarchical approach, I shall take a more holistic approach towards economic development which attaches as much importance to leakage prevention as to external income generation. In this approach, the creation of denser inter-linkages and synergies between all the sectors in an economy becomes just as important as generating external income. Therefore, the purpose of this book is not only to uncover the positive contributions of consumer services to local economic development but also, through the study of consumer services, to reconceptualize economic development theory and practice.

THE STRUCTURE OF THE BOOK

To start to move towards this retheorization of the role of consumer services in particular, and economic development theory and practice in general, Part I of this book commences by evaluating manufacturing (chapter 2), producer services (chapter 3) and consumer services (chapter 4) as motors for local economic development. In each case, their size, growth and location as well as their contributions to economic development are analysed. This reveals that although deindustrialization in the advanced economies has led to a rethinking of the role of services in regenerating local economies (see chapter 2), for the most part, the focus has been upon producer services (see chapter 3). Consumer services, meanwhile, have received relatively scant attention. This is because even though they constitute the vast bulk of the service sector, consumer services are predominantly viewed as residual activities dependent upon other economic sectors for their vitality and viability. As chapter 4 shows, however, there are grounds for thinking that they play a more positive role in local economic development.

To explore these contributions, Part II examines the roles that specific consumer service industries play in local economic development. Here, five industries are analysed, all chosen because they make major contributions to local economic development and because, in consequence, a large enough body of evidence is available to evaluate their economic impacts.[4] These are tourism (chapter 5), sport (chapter 6), universities (chapter 7), retailing (chapter 8) and the cultural industries (chapter 9). All these chapters adopt the same structure. First, they uncover the magnitude, character and spatial distribution of the industry, showing how this is the product of more than simply local demand. Second, they document the specific nature of their contributions to local economic development both as external income generators and leakage preventers and thus in terms of how they inter-relate with other sectors. Third, they make suggestions about the changes required in local economic policy to enable these contributions to be improved.

Following this, Part III undertakes five British locality studies to provide an illustrative dimension of the specific contributions of consumer services in different geographical contexts. These cover the second-tier urban economies of Glasgow (chapter 10), Sheffield (chapter 11) and Leeds (chapter 12), the Fens rural economy in East Anglia (chapter 13), and finally, a global city in the form of London (chapter 14). In each case, first, the economic restructuring of the locality is explored, second, whether consumer services have taken a lead or supportive role in local economic development is evaluated, and third and finally, their effectiveness in promoting economic regeneration is analysed.

After a review of the contributions of specific consumer services in Part II and an illustration of their contributions in particular localities in Part III, chapter 15, by way of conclusion, brings together these findings to summarize both the positive contributions of consumer services to local economic development and the implications for economic development theory and practice of this analysis. This will reveal that the study of consumer services, far from being of little value as many service-sector researchers have often assumed, results in a fundamental reconceptualization of both the role of services in particular and economic development theory and practice more generally.

Part I

MOTORS OF LOCAL ECONOMIC DEVELOPMENT: A GLOBAL PERSPECTIVE

2

DOES MANUFACTURING MATTER?

Deindustrialisation and tertiarization in the global economy

INTRODUCTION

The notion that some economic activities function as 'motors' of local economic development whilst others do not derives from economic (or export) base theory (Glickman 1977, Haggett *et al.* 1977, Wilson 1974). To reiterate, grounded in the assumption that an economy needs to earn external income in order to grow, this theory divides any economy into two sectors: 'basic' activities, which generate external income and are seen as the 'motor' of economic development; and 'dependent' activities, which merely circulate income within localities and are perceived as 'residual' or 'parasitic' activities contributing little, if anything, to the economy.[1]

Based on this conceptualization, much debate has taken place concerning which economic activities are 'engines of growth' and which are 'parasitic'. The dominant way of categorizing activity in traditional discourse has been to perceive manufacturing as the basic sector and services as residual or dependent activities. Indeed, this view is by no means defunct. As Bachtler and Davies (1989: 168) point out: 'the view that the economy is solely manufacturing-driven still has widespread currency. This view is that services are wholly dependent on manufacturing and that service jobs are not "real" jobs.' It is frequently prevalent in the economic development strategies of local government, in statements such as 'most services in the district are of an essentially local nature' (Kirklees Metropolitan Borough Council 1991:6). Manufacturing 'die hards' are also to be found in the groves of academe. Campbell (1996: 49), for example, claims that '"service activities" depend on [manufacturing] for their survival and growth', whilst Peck and Tickell (1991: 36) argue that 'service industries are essentially

13

"parasitic" in that they do not actually add to wealth in the economy, although they can help to realise the value of wealth created elsewhere'. Erickson (1989), meanwhile, when examining the relationship between export performance and state industrial growth, surveys only the manufacturing sector. Services are ignored.

During the past decade, however, this simplistic dominant/subservient conceptualization of the relationship between manufacturing and services has come under increasing attack, not least because of the rapid and severe restructuring of many advanced economies. The aim of this chapter, therefore, is to evaluate critically what constitutes the basic sector in the contemporary world economy. Is manufacturing the sole motor of development or are service-sector industries also engines of growth? Indeed, are services replacing manufacturing as the principal motor of local economic development? Or does manufacturing still matter? To answer these questions, in the next section, the extent of deindustrialization and tertiarization in the global economy will be examined so as to help explain why there has recently been a reconsideration of which activities comprise the basic sector. This will be followed by a critical evaluation of the conventional categorization of manufacturing as 'basic' and services as 'dependent' activity and an assessment of the relative importance of manufacturing and services as engines of growth. Finally, conclusions will be drawn about whether manufacturing matters. This will reveal that although manufacturing has suffered severe contraction and services have rapidly grown, and despite services displaying an ability to export, manufacturing is still important. Not only does it remain the major source of exports for most localities in advanced economies, but the vitality of services remains inter-dependent (not dependent) on the continuing existence of a manufacturing base. The implication, therefore, is that although this book focuses upon consumer services, it does not deny the continuing importance of manufacturing activity in regenerating local economies.

DEINDUSTRIALIZATION AND TERTIARIZATION IN THE GLOBAL ECONOMY

Popular wisdom is that economies evolve through a series of stages in which first, primary industry, second, manufacturing and then services become the leading and most sizeable sectors. Labelled

14

the Fisher–Clark thesis (Clark 1940, Fisher 1935), such a view of the evolution of economies has been heavily criticized. As Table 2.1 displays, this model does not neatly fit the actual growth patterns in many national economies. The problem is that manufacturing has never dominated employment in most countries. In France, for example, the shift was out of primary industry and into services, whilst in nations such as the Netherlands, Japan and the USA, service-sector employment has always been higher than manufacturing employment. Indeed, in North America, service employment has been proportionately larger than manufacturing for all of this century (Bluestone and Harrison 1982). As Urry (1987) thus concludes, the Fisher–Clark model of a natural progression from agriculture through manufacturing to services as the largest sectors is at best specific to western Europe in the earlier years of this century.

However, even if most economies have never been dominated by manufacturing-sector employment and the Fisher–Clark staged model of development is ill-conceived, it cannot be denied that during the past three decades or so, tertiarization has been a global phenomenon. Between 1960 and 1990, the service sector's share of global output rose from 50.6 per cent to 62.4 per cent (see Table 2.2), and its share of world employment increased from 24.6 per cent to 35 per cent between 1965 and 1989/91 (see Table 2.3). This process, moreover, appears to be a near universal phenomenon. As Tables 2.2 and 2.3 show, although tertiarization is relatively slower in developing nations and the service sector remains proportionately smaller, there has been a global shift towards services throughout both the developed and developing nations of the world economy.[2]

Deindustrialization, however, is not so widespread. It is specific to the developed economies. During the past three decades, whilst developing nations have witnessed manufacturing growth (in terms of both absolute and relative output and employment), in the so-called 'industrial' nations, there has been a significant decline in manufacturing's share of total output and employment. The consequence is that manufacturing's share of output is now remarkably similar in both the developed and developing nations, although the trend is in opposite directions: declining in the developed nations and rising in the developing countries.

In major part, this is because the old international division of labour based upon sectoral differentiation (for example primary-

Table 2.1 Sectoral shifts in employment in industrial economies, 1970–84 (%)

	1870			1960			1984		
	Agriculture	Industry	Services	Agriculture	Industry	Services	Agriculture	Industry	Services
France	49	28	23	21	36	43	8	32	60
Germany	50	29	22	14	48	38	5	42	53
Japan	73	n/a	n/a	33	30	37	9	34	57
Netherlands	37	29	34	11	41	48	5	28	67
Sweden	54	n/a	n/a	15	42	43	5	29	66
UK	23	42	35	5	46	49	3	32	65
USA	50	24	26	8	31	61	3	25	72
Average	42	30	27	15	39	46	5	32	63

Source: Elfring (1989: Table 1)

Table 2.2 Structure of world output (% of GDP), 1960–90

	World Industrialized nations				Developing nations			
	All	*US*	*Japan*	*Europe*	*All*	*Latin America and Caribbean*	*Africa*	*Asia*
Agriculture								
1960	6.3	4.0	13.1	8.8	31.6	16.5	45.8	38.0
1970	3.9	2.8	6.1	4.8	22.4	11.9	33.1	23.9
1980	3.4	2.6	3.7	3.6	14.7	9.5	25.1	17.8
1990	2.5	1.7	2.4	3.0	14.6	8.5	30.2	17.4
Manufacturing								
1960	31.0	29.0	35.1	34.7	15.6	21.1	5.8	14.4
1970	27.7	25.2	36.0	30.5	17.8	22.8	8.6	15.1
1980	23.6	21.8	29.2	24.0	17.6	23.5	7.7	15.4
1990	21.5	18.5	28.9	21.5	20.7	22.3	10.9	20.9
Services								
1960	51.6	57.2	40.9	43.8	42.8	50.6	37.5	37.1
1970	58.3	62.7	47.2	54.3	40.1	55.3	44.0	42.5
1980	60.2	63.8	54.4	59.2	44.2	54.5	39.9	37.9
1990	64.7	70.3	55.8	63.8	49.8	57.5	41.6	45.9

Source: International Labour Organization (1995: Table 3)

Table 2.3 Sectoral distribution of world employment, 1965–89/91

	World	*Industrialized nations*	*Developing nations*
Agriculture			
1965	57	22	72
1989–91	48	7	61
Industry			
1965	19	37	11
1989–91	17	26	14
Services			
1965	24	41	17
1989–91	35	67	25

Source: International Labour Organization (1995: Table 4)

sector activity in developing nations and higher value-adding stages of manufacturing in developed nations) has been replaced by a new international division of labour (NIDL) based on the separation of functions. In this NIDL, the control and command functions are located in a network of global cities in the developed nations whilst physical production is increasingly dispersed into a host of developing countries where new and efficient technology can be allied to lower labour costs (for example Cohen 1981, Dicken 1992, Sassen 1991). During the 1970s and 1980s, the beneficiaries were the middle-income nations such as the Republic of Korea, Hong Kong and Singapore, who have now joined the ranks of the developed nations and been replaced by a new tranche of middle-income countries including Malaysia, Thailand and Indonesia (Hall 1996). The implications of this global economic restructuring for local economic development in Britain will be returned to in future chapters. For the moment, the important point to note is that whilst developing economies have witnessed both industrialization and tertiarization, the experience in developed economies has been deindustrialization coupled with tertiarization.

SERVICES: AN ENGINE OF GROWTH?

These twin trends of deindustrialization and tertiarization in the developed nations and the accompanying NIDL have prompted a major rethinking of the traditional view of services in both neo-classical economics and marxism as 'unproductive' and 'parasitical'. Here, therefore, the conventional categorization of manufacturing as 'basic' and services as 'dependent' activity will be critically evaluated and the relative importance of manufacturing and services as engines of growth will be assessed. To do this, first, the magnitude and character of international trade in services will be examined, then the extent and nature of inter-regional and extra-local trade in services will be analysed. Subsequently, a case study of Canada will be employed to assess in greater detail the relative importance of manufacturing and services as engines of growth.

International trade in services

To assess the significance of international service exports, commentators have attempted to estimate the proportion of total international trade composed of services. Daniels (1993), for example, estimates that in 1986, world receipts by major balance of payments components amounted to more than $3 trillion, of which more than $1 trillion comprised trade in services. Segebarth (1990), meanwhile, finds that almost 22 per cent of total international European Union (EU) trade is accounted for by services and 16 per cent of total international US trade. Moulaert and Todtling (1995b), furthermore, explore the relative magnitude of international service exports and whether their share of total exports is increasing or decreasing over time, using OECD data which are reproduced in Table 2.4. They find that services represent between one-quarter and one-fifth of total exports on average in the advanced economies. Far from being 'dependent' activities, therefore, they are a major component in their export base.

Contrary to popular prejudice, however, the share of manufacturing trade in total exports is not decreasing. Quite the opposite. As the International Labour Organization (1995) reports, the manufacturing sector's contribution to total global exports increased from 60.9 per cent in 1970 to 71.1 per cent in 1990, and from 72.0 to 78.0 per cent in advanced economies. This is reinforced by Table 2.4, which shows that between 1970 and 1991, the share of service trade in total exports decreased by over 3 percentage points on average. Services, therefore, are not an ever greater proportion of total international exports in these developed economies. Nevertheless, this overall trend masks major differences between nations. Some advanced economies such as Denmark, Finland, France, Sweden and the USA have witnessed an increase in the share of services in total exports. Others, however, such as

19

Table 2.4 Trade in services for different OECD countries: in $million
(current prices)

	% of services in total exports			Exports of services as % of GNP		
	1970	1980	1991	1970	1980	1991
Norway	46.47	31.24	28.24	20.71	24.86	18.71
Iceland	37.92	30.75	28.98	18.20	17.57	13.91
Portugal	34.83	31.06	25.68	4.24	5.28	6.24
Austria	29.84	14.63	11.86	5.88	6.20	5.77
Italy	27.35	18.50	21.52	3.01	3.84	4.97
Britain	27.19	23.76	22.29	4.27	7.71	5.89
Denmark	25.94	26.22	32.70	6.54	12.70	17.44
France	22.56	29.42	29.51	2.84	8.30	7.74
USA	21.72	15.32	24.72	1.27	1.57	2.60
Sweden	18.75	19.48	21.73	4.87	9.37	9.99
Germany	17.62	16.89	13.05	3.08	6.04	5.58
Canada	16.29	12.45	14.03	4.06	3.78	3.88
Finland	16.07	17.04	19.30	3.44	7.61	6.71
Japan	14.73	13.81	12.95	1.11	2.14	1.92
Average	25.52	21.47	21.90	5.96	8.36	7.95

Source: OECD statistics on international transactions (in Moulaert and Todtling
1995b: Table 10.2)

Austria, Germany, Britain, Italy, Norway and Portugal, have seen a
decline in the percentage of total exports derived from services.
Consequently, there are wide variations between developed nations
in both the extent to which services are an engine of growth as
traditionally defined and according to whether the relative importance of services is increasing or decreasing over time.

So too are there major differences in how the international trade
in services is distributed between nations. As Daniels (1993)
reveals, most international trade in services takes place to and
from a small number of nations. Almost 81 per cent of all service
exports originate from just 20 nations (mostly developed market
economies), and the same 20 nations receive 77 per cent of total
world imports. Indeed, just five countries generate 44 per cent of
the exports and 45 per cent of the imports (US, France, UK,
Germany and Japan). At the pinnacle is the US, with 12.1 per
cent of total service exports and 11.5 per cent of total service
imports. Consequently, 'There appears to be a relationship between
the level of economic development of a country and the propensity
to export services or to generate demand for service imports'
(Daniels 1993: 83).

Not only are developed nations the major participants in international service trade, but such trade plays an important role both in generating external income for these economies and in determining whether their net balance of payments is positive or negative. In developing nations, service imports exceed exports by a margin of some 34 per cent, whilst in advanced economies such as France and the UK, service surpluses are normal and make up for the deficits on goods trade (Daniels 1993). Therefore, service exports are an important determinant in ensuring that such countries increase their net income and continue to grow.

Inter-regional and inter-local trade in services

If trade in services represents a large component of the export base in many developed nations, the same is true of local and regional economies. Again, however, such activity is unevenly distributed. Examining the 'location quotients' (LQs) of services in different regions and localities to determine their relative specialization in services (where LQs of greater and less than 1.0 signify an over- and under-representation of services respectively), Marshall *et al.* (1987) find that core regions in Britain had service-sector quotients of between 0.99 and 1.19 whilst peripheral regions had service location quotients of 0.40–0.69. They conclude that core regions dominate the national system of service trade. In North America, Keil and Mack (1986) and Gilmer *et al.* (1987) identify similar trends.

To assess both the extent to which services export inter-locally and inter-regionally and the propensity of services to export relative to manufacturing, a large number of empirical studies have been conducted on specific localities, many of which are analysed in the next chapter. Here, however, and to provide an indication of the relative export propensity of services, a selection is briefly reviewed. Porterfield and Pulver (1991), for example, examine manufacturing and services in the Upper Midwest states of the US. They find that 44 per cent of manufacturers' sales and 32 per cent of services' sales were inter-local (i.e. beyond a 50-mile radius of the establishment), whilst 20 per cent of manufacturers' sales and 12 per cent of services' sales were inter-regional (i.e. outside the Upper Midwest region). The evidence, therefore, is that although services export and are thus not dependent activities, manufacturing is more export-oriented than services. Many other locality studies arrive at the same conclusion (for example Beyers and Alvine 1985, Stabler and Howe 1993, Williams 1996a).

Nevertheless, when the manufacturing and service sectors are broken down into their various sub-sectors for a finer-grained analysis of the proclivities of different manufacturing and service industries to export, it is revealed that manufacturing industries do not always display a greater export propensity than services. In a survey of the rural Fens economy, for example, Williams (1996a) finds that although manufacturing displays a higher export orientation than the service sector, producer services have the same export propensity as the metal goods, engineering and vehicle-manufacturing industries. In a study of 1500 service firms in New England, moreover, Ashton and Sternel (1978) discover that whilst 37 per cent of services derive at least 10 per cent of their revenues from clients in the US but outside New England, three-quarters of R & D and management consultancy businesses obtain at least 10 per cent of their revenues from this external source, whilst the proportions are much lower in equipment leasing (34 per cent) and advertising (28 per cent). Other studies, however, identify different sub-sectors as having the greatest propensity to export. Beyers and Alvine (1985), for instance, deem transport services and business and financial services sectors to have substantial extra-regional revenues whilst Van Dinteren (1987) identifies computer services and advertising services as being most export-oriented. These sub-sectoral variations in the proclivity of services to export will be returned to in the next chapter. Here, it should simply be noted that although manufacturing as a whole has a higher export orientation than services, this does not necessarily apply when these sectors are disaggregated into their component industries.

Export propensity, furthermore, varies not only across sub-sectors but also according to firm size. In the same New England survey cited above, Ashton and Sternel (1978) find that small firms (annual sales under $500000) displayed less propensity to export than larger firms ($20 million of sales annually): 27 per cent and 64 per cent respectively received over 10 per cent of revenues extra-regionally. So too were externally owned companies more likely to export than indigenous companies. Other studies identify the same relationship between export propensity and firm size and ownership (Marshall 1983, Smith 1984, Van Dinteren 1987). However, such findings are not uniform across all studies. Beyers and Alvine

(1985) and Michalak and Fairbairn (1993), for example, identify no relationship between firm size and export orientation.[3]

Several reasons exist for these variations. In major part, it is because of the contrasting nature of the localities and regions studied. Different local economies will vary in terms of the industries which display a propensity to export and the extent to which they do so. The consequence, therefore, is that generalizations about the proportion of output sold externally by any given industry cannot be made from such one-off locality studies. Neither, moreover, can their results be compared, since both the definition of 'local' and the survey methods employed are often very different across studies.[4] Nevertheless, even if such empirical studies are not directly comparable, their significance is that they provide a wealth of evidence that services act as 'basic' activities in many local and regional economies. Despite this, very few identify the relative importance of manufacturing and services as engines of growth at a local and regional level, and even fewer explore how this is changing over time. For this reason, the next section turns towards a case study of Canada.

Are services replacing manufacturing as the engine of growth? The case of Canada

Unlike most other surveys, the input/output data produced by Statistics Canada allow detailed analysis of the relative importance of services and goods production as engines of growth as well as of how this is changing over time. As Table 2.5 reveals, 24 per cent of all goods production was exported internationally in 1974 (28 per cent in 1984), compared with just 6 per cent of all service production (4 per cent in 1984). Consequently, whilst services represented 17.6 per cent of total foreign exports in 1974, this had fallen to 12 per cent by 1984.

Nevertheless, at an inter-regional level, 45 per cent of all goods production was exported outside the province of origin in 1974 (51 per cent in 1984), compared with 43 per cent of all service production (42 per cent in 1984). Hence, whilst 44.9 per cent of all provincial external-income generation came from service exports in 1974, this had slightly fallen to 44.1 per cent by 1984. So, the conceptualization of the service sector as a dependent locally oriented industry is far from the truth in Canada. Although the export orientation of service production is slightly lower than

Table 2.5 Destination of provincially produced goods and services, 1974 and 1984, Canada

% of production	Goods		Services	
	1974	1984	1974	1984
Exports				
Other Canadian provinces	22	23	37	38
Rest of world	24	28	6	4
Total	45	51	43	42
Within province to Intermediate:				
Service-producing industries	5	5	21	19
Goods-producing industries	22	20	17	15
Investment expenditures	16	16	2	2
Total	43	40	39	36
Final demand:				
Consumers	11	9	15	17
Governments	1	0	3	5
Total	11	9	17	22
Total ($m 1984)	299220	359024	255747	344020

Source: Derived from Stabler and Howe (1993: Tables 1 and 2)

for goods production, it still provides just under half of total external income generation.

For some analysts, such data on the export propensity of services are argued to be due to a misallocation of 'goods production' to the so-called service sector (Sayer and Walker 1992). This is not the case. Using this input/output data from Statistics Canada, which is one of the few data sources available on this issue, Stabler and Howe (1993: Tables 5 and 6) examine the value of the goods manufactured in service-producing industries and services generated in goods-producing industries. They reveal that in 1984, 1.8 per cent of total production by value was misallocated to the service-producing industries. Since service-producing industries increased their share of total production in Canada from 46 per cent to 51 per cent between 1974 and 1984 (Stabler and Howe 1993: Tables 5 and 6), great care should be taken not to over-emphasize this concern about a misallocation of goods production. Service-producing industries and services are a rapidly growing share of the economy, and this is not merely due to a misallocation of goods production to such industries.

24

DOES MANUFACTURING MATTER?

Indeed, there is evidence that the above data on the tradability of manufacturing and services, if anything, under-emphasize the contribution of services to external-income generation. This is because they fail to account for services embodied in goods exports and vice versa, such as in intra-corporate exports where the corporate or divisional headquarters of a multi-facility company provides services or goods to the entire company or division. Yet intra-corporate flows represent a major element of exports and imports. Polese (1982: 158), for example, finds in eastern Quebec that almost half of service demand is satisfied by imports and that 'almost 45 per cent of the service imports are in the form of intra-firm flows'.

Consequently, when these embodied exports of goods and services are added to direct sales, the relative importance of services in total exports is substantially increased. As Table 2.6 reveals, if direct sales only are examined, service exports constituted 34 per cent of value-added in total exports in 1984, but 46 per cent when embodied exports are incorporated. More importantly, when embodied exports are taken into account, the relative importance of services is increasing over time. Services increased their share of value-added in total exports from 41 per cent to 46 per cent between 1974 and 1984, supporting the view that the Canadian economy is increasingly dependent on service inputs and service exports.

In sum, the conventional categorization of manufacturing as 'basic' and services as 'dependent' activity is far from the reality in contemporary local economies. Services act as engines of growth in many localities through their ability to trade inter-locally, inter-regionally and internationally. Indeed, the evidence from

Table 2.6 Goods and services share of value-added in exports 1974 and 1984, Canada

% of value-added in exports	Goods exports	Service exports
Direct exports		
1974	70	30
1984	66	34
Direct plus embodied exports		
1974	59	41
1984	54	46

Source: Derived from Stabler and Howe (1993: Tables 7 and 8)

Canada suggests that when embodied exports are taken into account, the relative importance of both service inputs and exports is increasing over time.

DOES MANUFACTURING MATTER?

This chapter has thus shown that although manufacturing in the advanced economies has suffered severe contraction and services have rapidly grown, and despite services displaying an ability to export, it would be erroneous to write off the manufacturing sector. Manufacturing still matters. It contributes a large share of total external income in most local economies and its propensity to export remains, on the whole, higher than that of the service sector. Consequently, it is not the case, as some service-sector researchers have commented, that the service sector is now the principal 'engine of growth' (Riddle 1986, Swan 1985). To argue this is to ignore the evidence. Although the share of services in total exports may be increasing, albeit only when embodied exports are taken into account, manufacturing and goods exports still provide the vast majority of exports in most local economies.

Moreover, to argue that services are now the engine of growth is also to ignore the complexities of modern economies in terms of how manufacturing and services inter-relate. It is not the case that services are reliant on manufacturing for their growth, but neither is it the case that services and manufacturing are independent of each other. Rather, there is an inter-dependency. As Marshall and Wood (1995: 6) assert,

> Services offer specialist skills to society. Their value, therefore, depends upon how they are used elsewhere, in primary, manufacturing or other service production, or by consumers. In the past, this inter-dependence has sometimes been mistaken for dependence. It should now be widely recognised that service functions are vital to both the effectiveness of production and the quality of life.

It is important, therefore, not to fall into the same trap as some of the early service-sector commentators of attempting to displace the previously dominant position of manufacturing in economic development with a service-dominant conviction (for example Riddle 1986, Swan 1985). Instead, and as Marshall and Wood (1995) assert, there is a need to recognize the complex interactions between goods- and service-producing activities in forms of production which are increasingly reliant upon service expertise. The lesson, therefore, is that local economies cannot and should not seek economic growth solely through either manufacturing, producer services or consumer services since each is mutually dependent on the other. Indeed, each is equally capable of initiating economic change and each will play different roles in different economic and geographical circumstances.

To start to tease out some of the complexities involved in understanding the contexts in which services may lead or follow economic development, the next chapter explores the role of producer services, a sector upon which much has been recently written and which has come to dominate the study of the service sector.

3

PRODUCER SERVICES AND ECONOMIC REGENERATION

INTRODUCTION

The aim of this chapter is to explore the contributions of producer services in general, and advanced producer services (APS) in particular, to economic regeneration in developed nations. Producer services are intermediate functions that serve as inputs into the production of goods or other services, whilst APS, upon which much of the recent literature on producer services concentrates, are more specifically those 'complex knowledge-intensive business services designed as direct inputs to firms' (Moulaert and Todtling 1995a: 102). Here, therefore, and through a review of the vast literature on producer services and APS which has emerged since the 1980s, this chapter will reveal some of the complexities involved in understanding the interactive dynamic relationships between such services and the rest of the economy. First, the size and growth of producer services and APS will be examined, second, the location of these services will be explored, and third and finally, their contributions to economic development will be analysed.

Before commencing, however, it is important to note that the definitions of producer services and APS are not homogeneous across studies. This is not only because whether a service caters mostly to intermediate or final demand varies temporally and according to geographical location, but also because some studies classify activities which mostly meet final demand as producer services anyway (for example banking).[1] The problem is perhaps that as 'producer services' have been recognized to play a positive role in economic development, some activities which are essentially consumer services but display a proclivity to export have been slotted into the producer services category to provide them with heightened status. The consumer services grouping, meanwhile, much like the service sector as a whole in the past, has become the 'residual' category into which supposedly 'parasitic' activities have been placed. The reader, therefore, should keep in mind that some of the activities considered here will be in certain times and places essentially consumer services.

SIZE AND GROWTH OF PRODUCER SERVICES

Producer services, however defined, are not new. Activities such as transport, distribution, legal, architectural, financial and insurance services have existed since the first days of the exchange economy, whilst others developed with industrialization. They have come to prominence during the past few decades because of their absolute and relative growth. Examining England and Wales, Daniels (1995a) argues that up to the mid-1950s, producer services remained relatively stable at about 3 per cent of total employment. As economic growth recovered after the Second World War and the economy began to participate in world markets, producer services began to flourish. Their share of total employment had reached 4 per cent by 1960, 5 per cent by 1970, 8.6 per cent by 1981 and 10.8 per cent by 1991. Indeed, whilst total employment in Britain rose by just 2 per cent between 1981 and 1991, producer service jobs increased by 28.4 per cent.[2] Similarly rapid growth rates have been documented in both Canada (Coffey 1995a) and the USA (Harrington 1995b).

Not all producer services, however, are expanding at the same rate. In Britain between 1981 and 1991, according to the Census of Employment, business service employment expanded by 125.7 per cent, auxiliary services to banking and insurance by 61.4 per cent, professional and technical services and advertising by 38.2 per cent, and wholesale distribution by just 6.3 per cent. Other producer services, however, such as employment in research and development, declined by 25.6 per cent. These patterns of expansion and contraction are by no means unique to Britain. In most countries, it has been the information, knowledge- and innovation-related services (for example management and organization consulting, technical services, computer and software services) as well as services related to marketing and distribution (for example market research and consultancy, advertising) which have grown quickest (Moulaert and Todtling 1995b). Indeed, it is for this reason that so much attention has been placed upon APS.

To document the patterns of growth in APS, Moulaert and Todtling (1995b) analyse employment change in finance, insurance and real estate (FIRE) and business services in OECD nations. They find that these grew little until the late 1960s but then expanded rapidly throughout the 1970s and 1980s so that by 1990, shares of APS in national employment in these developed economies ranged from 4 to 12 per cent (see Table 3.1). However, the period in which APS underwent rapid expansion, not to mention their growth rates, varied significantly between nations. In the more advanced economies (UK, France, Germany, Nordic countries), the APS sector developed earlier than in either 'intermediate' countries (Austria, Italy) or less developed advanced economies (Portugal, the former East Germany) where APS growth has occurred only since the 1980s. Growth rates, meanwhile, are of three types so far as western Europe is concerned: fast (France, UK), intermediate (Italy) and slow (Germany and Switzerland), reflecting national variations in government policy towards manufacturing, capital availability and levels of investment in different sectors (Bailly 1995). Indeed, saturation appears to have been

Table 3.1 Percentage of civilian population employed in finance, insurance, real estate and business services, 1970–90

	1970	1980	1990
Denmark	5.9[a]	6.6[b]	9.4
Austria	3.4	5.0	6.5
Finland	3.5[a]	5.5	8.3
France	5.3	7.5	10.0
Germany	—	5.8	7.9
Britain	5.0	7.3	10.4[e]
Italy	—	2.6	4.2
Iceland	4.0	5.4	8.2[c]
Norway	4.0[d]	5.4	7.5
Portugal	1.9	2.0	4.6
Sweden	5.0	6.7	8.6
USA	6.8	8.4	11.3
Japan	2.6	5.7	8.3
Canada	4.8	9.5	11.6

Source: OECD Labour Force Statistics, Paris 1992 (cited in Moulaert and Todtling 1995b)
Notes: a = 1971; b =1981; c =1989; d =1972; e =1987

reached in some economies (USA, Britain, France and Sweden), with growth rates falling or coming to a halt in the late 1980s in the UK and Nordic economies (Moulaert and Todtling 1995b).

LOCATION OF PRODUCER SERVICES

The dynamics of producer service and APS growth differ not only cross-nationally but also intra-nationally. To understand the location of producer services within countries and how this is changing over time, a useful starting point is the NIDL thesis discussed in the last chapter. This holds that headquarters are concentrated in global cities that serve as centres of corporate control and administration in transnational capitalism (Cohen 1981, Noyelle 1984, Sassen 1991). Divisional headquarters, meanwhile, may be located in second- or third-tier urban centres, while the production branch plants are very often located in traditional industrial areas or in industrialising rural areas. Although this is more usually applied to manufacturing industry, when it is related to producer services and APS, a similar spatial division of labour is uncovered. Not only is producer service and APS employment, as will now be shown, concentrated in the major agglomerations (Marshall 1979, Noyelle 1983, Daniels 1995a, Cappelin 1989), but so too are the control functions of larger producer-service and APS firms (Howells 1988, Daniels 1995a and b, Moulaert, Todtling and Schamp 1995), which deliver highly specialized management and marketing consulting, accounting/auditing, financial and R & D services.

Examining the concentration of APS in capital cities, Moulaert and Todtling (1995b) find that Oslo and Copenhagen had half of total national APS employment in 1981, Vienna 46 per cent, London 43 per cent, the Paris region 41 per cent and Stockholm 40 per cent, all well above their share of total national employment. In the former West Germany, the situation is somewhat more decentralized with a larger number of agglomerations having high shares, although they remain concentrated in large cities (Schamp 1995). The same holds for Italy, where in 1992, the north-west of the nation had 30.9 per cent of all APS suppliers, the north-east 41.7 per cent and the peripheral south just 26.4 per cent (Moulaert and Todtling 1995b). The strongest spatial concentrations, however, are in those countries which have only relatively recently developed a robust producer services sector. In Portugal in 1981, for instance, 56 per cent of producer service jobs were in

Lisbon and another 18 per cent in Oporto (Farrao and Domingues 1995).

Nevertheless, capital cities lost some of their dominance over the 1970s and 1980s. The share of APS employment in Lisbon declined from 61 per cent in 1970 to 56 per cent in 1981, whilst in London, it declined from 49 per cent in 1971 to 43 per cent in 1981, in the Paris region from 44 per cent in 1975 to 39 per cent in 1990, and in Vienna from 49 per cent in 1976 to 44 per cent in 1988 (Moulaert and Todtling 1995b). In general, however, such decentralization does not imply any fundamental change in the spatial pattern of producer services. There is still a strong over-representation in the largest agglomerations (Beyers 1993, Daniels 1995a and b, Todtling and Traxler 1995, Moulaert and Gallouj 1995, Moulaert and Todtling 1995b).

Moreover, it is not only producer service employment which is concentrated in these principal cities but also the most advanced, specialized and high-level producer service and APS firms as well as the control functions of the largest companies (Coe 1996, Illeris and Sjoholt 1995, Lundmark 1995, Moulaert and Gallouj 1995, Moulaert and Todtling 1995b, Todtling 1984, Todtling and Traxler 1995). In Austria, for instance, 76 of the 100 largest APS firms are located in Vienna (Todtling and Traxler 1995), whilst in the UK, some 56 per cent of the business service firms employing 500–1000 staff and almost 70 per cent of those employing more than 1000 are headquartered in London, as are 48 per cent of those with turnover in the range £1–5 million and 57 per cent of those with a turnover exceeding £5 million (Daniels 1995a).

Consequently, where producer services and APS have been decentralized, this has mostly been a case of the routine functions, less advanced producer services and smaller firms supplying regional and local markets (Illeris and Sjoholt 1995, Moulaert and Todtling 1995b, Todtling 1984, Todtling and Traxler 1995). For example, back offices may seek cheaper labour and lower property costs through decentralization, or franchises and local offices may be set up to serve a region. This model of geographical dispersion, therefore, parallels the developments in manufacturing, both in terms of the resulting spatial divisions of labour and in terms of the limited local economic impacts of the branches, franchises and sub-contracted activity (for example Foley et al. 1996).

This dispersion process, moreover, is not occurring evenly. It is a concentrated decentralization. In the Nordic countries, Illeris and

Sjoholt (1995) identify that producer services have not spread to peripheral less dynamic regions, as does Daniels (1995a and b) in the UK, who finds that although there is some intra-regional redistribution from the largest cities to adjacent towns, the inter-regional distribution of producer services remains heavily concentrated in the affluent southern core of the country, especially of higher-order business services (Daniels 1995a and b, Howells 1988). Similar concentrated decentralization trends have been identified in Austria (Todtling and Traxler 1995) and France (Moulaert and Gallouj 1995).

In seeking to explain why such producer services and APS remain clustered in capital city regions, most commentators agree on the major reasons, even if there is some debate on which is the most important in different contexts. These include: accessibility and proximity to each other; good physical access to customers and a large range of other local business activities; readily accessible transport facilities; a competitive market environment, maintaining the quality of services and providing the basis for exports elsewhere; the availability of high-quality telecommunications infrastructure; high-quality labour; large numbers of clerical and administrative staff; sites accessible and attractive to staff; an ample supply of suitable-quality office accommodation; and a high-quality urban environment, including cultural, social and shopping facilities (Marshall and Wood 1995).

Besides concentrated decentralization, a further spatial trend is that there is increasing regional and local specialization in producer services and APS. Examples include financial services in London, textile engineering consulting in Lille-Roubaix-Tourcoing, services related to high-tech firms in Rhône-Alpes and the south of France, engineering and software in Munich, accounting and consultancy in Frankfurt, advertising in Hamburg and Düsseldorf, certification of ships in Oslo and environmental services in Copenhagen (Moulaert and Todtling 1995b). In these local economies, the relationship seems to be two-fold: on the one hand, APS activities gain advantages of specialization and competitive strength from sophisticated nearby customers and partners; and on the other hand, the custom-tailored APS sector improves the competitiveness of the respective regional economies (Moulaert and Todtling 1995b).

As Moulaert et al. (1995) reveal, this local specialization can take the form of either intra-urban spreading (for example multi-polar cities with a number of central business districts), an escape to the hinterland of metropolitan areas or an increasing presence in non-central regions. However, there are important qualifiers referring to the nature of the producer services involved and the areas concerned. First, it is usually not the peripheral regions which benefit from this specialization process but rather, regions near to or adjacent to major financial or corporate urban centres. Second, there are qualitative differences between APS implantation in large metropolitan regions and in provincial centres since the former harbour primarily headquarter functions and related high-level services, whilst the latter generally house branch plants or small firms engaged in less advanced producer services (Howells 1988, Leyshon et al. 1988). Where regional and local specialization occurs in producer services, therefore, it further reinforces rather than mitigates the spatial divisions of labour already prevalent.

In sum, although there is some diffusion from the metropolitan cores to the surrounding areas, peripheral regions tend to fall beyond the scope of APS activity. Moreover, it is mainly the routine and less advanced producer services (data processing, book-keeping, routine financial services) and small firms that are growing rapidly outside metropolitan areas, and even though there is some evidence of specialized concentration, most advanced producer services and the control functions of large firms continue to be concentrated in metropoles or high-technology manufacturing centres. Therefore, local and regional inequalities in producer service location adopt the form of increasing differentiation between high-level services which are concentrated in major urban agglomerations but have witnessed some 'concentrated decentralization' within the core regions (Moulaert and Todtling 1995b) and other lower-order less advanced producer services which follow a more even spatial distribution (Moulaert et al. 1995). This evolving geography of producer services and APS is important. As the next section reveals, their relative clustering in the larger metropolitan city regions leads to a reinforcement of uneven development because of the contributions which such services make to economic development.

THE ROLE OF PRODUCER SERVICES IN ECONOMIC DEVELOPMENT

The contributions of producer services to economic regeneration are argued to be that they export their products beyond their

34

boundaries and contribute to the competitiveness and productivity of other economic activity by enhancing the efficiency of operation and the value of output at various stages in the production process, including activities both upstream (for example research) and downstream (for example marketing) of actual production. Here, therefore, a brief summary of each is provided.

The export role of producer services

There is now a wealth of evidence that producer services export (Beyers and Alvine 1985, Coffey 1995b, Esparza and Krmenec 1994, Goe 1994, Harrington 1995a and b, Harrington *et al.* 1991, Moulaert and Todtling 1995b, O'Farrell 1993, O'Farrell and Moffatt 1995, O'Farrell *et al.* 1992, 1995). Examining producer services in New York City, for example, Drennan (1992) finds that two-thirds of all export earnings are generated by this sector. In a study of the Central Puget Sound region, meanwhile, Beyers and Alvine (1985) discover that producer services conduct over a third of their business outside the region, and that the employment resulting from service activities is as large as export-related employment in manufacturing. Given the similar economic structure of the Central Puget Sound to other large North American metropolitan areas, they suggest that this could be replicated elsewhere.

However, different producer services display varying propensities to export. For example, comparing the contributions of business services and the FIRE sector in Montreal, Coffey (1995b) reveals that although 18.9 per cent of gross revenues in the producer services sector as a whole derive from exports outside the local economy, this is 24.6 per cent for business services and 12.2 per cent for FIRE firms. Indeed, whilst producer services as a whole generate 7.9 per cent of gross revenues from extra-regional exports, business services receive 9.3 per cent and FIRE services 5.7 per cent. Similarly, and so far as international exports are concerned, whilst producer services derive 3.4 per cent of their revenues from such exports, this is 4.4 per cent for business and 1.8 per cent for FIRE services.

This tradability, moreover, varies by location. Examining the export propensity of various sub-sectors of business services in Ireland, Scotland and Nova Scotia, Table 3.2 reveals that in Scotland and Ireland, product design is the most export-oriented. In Nova Scotia, meanwhile, advertising and marketing have the greatest

Table 3.2 Export propensity of business services: Ireland, Scotland and Nova Scotia

	Ireland			Scotland			Nova Scotia		
	% of sales which are								
	local	ROI	intl	local	RUK	intl	local	RAC	intl
Advertising and marketing	76	10	14	58	40	2.0	36	44	2
Management consultancy	53	28	18	51	43	7.0	58	35	6
Market research	66	14	21	35	59	0.2	46	47	2
Graphic design	89	11	0	61	36	3.0	74	18	2
Product design	30	38	33	24	53	23.0	n/a	n/a	n/a
All business services	66	19	16	56	41	4.0	53	37	4

Source: Derived from O'Farrell *et al.* (1995) and O'Farrell (1993)
Note: Local is within a forty-mile radius; ROI, rest of Ireland; RUK, rest of United Kingdom; RAC, rest of Atlantic Canada; intl, international exports beyond the nation

propensity to export beyond the locality and management consultancy internationally. There are also significant cross-national variations in the proclivity of each of these producer services to export. Advertising and marketing firms, for example, export 24 per cent, 42 per cent and 64 per cent of their sales beyond the locality in Ireland, Scotland and Nova Scotia respectively.

The propensity to export, therefore, is dependent upon the nature of the economy inhabited. For example, and as O'Farrell *et al.* (1996) identify when comparing the south-east of England and Scotland, business services in global cities and their regions tend to display a greater export propensity than those in second-tier cities or peripheral regions, thus reinforcing the dominant position of such areas. Further supporting this view, Daniels (1995a) finds that business services in the City of London export a greater proportion of their output than business services in second-tier cities. Indeed, more than 21 per cent of the turnover of 50 business service firms located in the City of London in 1991 originated outside the UK, compared with just 7 per cent in Bristol, 3 per cent in Leeds, 8 per cent in Edinburgh and 13 per cent in Cardiff. The destination of such exports also differed. For City of London business services, non-EU markets (13 per cent of total sales) contributed more than EU markets (8 per cent). This is the inverse for business services located in other UK cities in that

EU markets are the more likely destination for exports. Second-tier cities, therefore, have a lower propensity to export than global cities. In major part, this is because these global cities offer the knowledge base as well as the transportation and communication facilities needed for international trade, finance and indirect investment (Daniels 1995a and b), whilst firms more oriented to regional markets are inclined to follow or stay next to their clients, for example by setting up branch offices in provincial cities (Coe 1996, Moulaert and Todtling 1995b). The tradability of producer services, nevertheless, is not their sole contribution to economic development. A further role lies in their ability to enhance the competitiveness and productivity of other sectors of the economy.

The relationship of producer services to other sectors

Through their role in investment, innovation and technological change, producer services make a key contribution to economic development by influencing the capacity of an economy to adjust to changed economic circumstances (Coffey 1995a). As the case study of Canada in chapter 2 demonstrated, service inputs are increasingly involved in the production process, for example in the form of design, research, consultancy, marketing, financial and computer services. Marshall and Wood (1995) explain this increased use of service inputs as being due to a number of factors. These include: the emergence of new goods and service products requiring specialist service support; transformations in how goods and services are produced arising from process innovations which increase specialist service demand; increasingly complex and internationally integrated financial, production and distribution environments which require additional service support; changes in government regulation and intervention, invariably requiring businesses to monitor and analyse changes; and the proliferation of tasks related to the internal management and administration of firms, especially complex multinational businesses.

It is important to be clear, however, that in stating that producer services are increasingly involved in overall production, the idea is not conveyed that such services merely provide a support function to the endeavours of other 'leading' economic sectors such as manufacturing. Neither is it being suggested that they are important solely through their ability to enhance productivity and efficiency in these other sectors, or that their rapid growth is simply due to the externalization or 'unbundling' of functions previously provided in-house by manufacturing firms.

It is now widely understood that producer services are consumed not only by manufacturing establishments but by firms across all sectors of the economy (Coffey 1995a, Daniels 1995a, Moulaert and Todtling 1995b, O'Farrell 1995). Goe (1994), for example, finds that only 5.4 per cent and 8.7 per cent of producer services in Cleveland and Akron respectively trade exclusively with manufacturing industry. Meanwhile, 57.0 per cent and 63.9 per cent of producer service establishments trade only with service-sector industries. Similarly, in a survey of 324 producer service firms in Montreal, Coffey (1995b) uncovers that just 5.3 per cent and 21.8 per cent of the gross revenues of FIRE and business service establishments respectively are derived from the goods-producing sector. In fact, 40.6 per cent of business service establishments receive absolutely no revenues from the goods-producing sector. Even in the manufacturing-dominated economy of Germany, manufacturing still only consumes 42.6 per cent of all producer service output (Schamp 1995) and 39 per cent in Austria (Todtling and Traxler 1995). The service sector itself, therefore, is widely acknowledged as the most important market for producer services. In consequence, producer services are not simply reliant on manufacturing.

Neither is the externalization of functions by manufacturing the sole explanation for the growth of producer services. Instead, and as Marshall and Wood (1992, 1995) amongst others recognize, any understanding of the role of producer services in contemporary economies must recognize the inter-dependencies between the various sectors of the economy. The challenge for those involved in the study of producer services, therefore, is increasingly one of charting the nature of these inter-dependencies between producer services and other sectors of the economy so as to shift thinking away from the logic of one-way dependencies such as that producer services are reliant upon manufacturing's externalization of service functions. The outcome, as both Illeris (1989a and b) and Marshall and Wood (1995) identify, is that producer services can no longer be studied in isolation from other sectors of the economy. So too is there an appreciation that these inter-relations are both temporally and spatially differentiated and vary according to such factors as firm type and size (for example Hansen 1994). Today, therefore, the study of producer services has moved far beyond simply investigating whether or not they directly contribute to exports. How their inter-relations with other sectors can be further developed to create local synergies, for example, has become an important issue on the research agenda.

CONCLUSIONS

Building upon chapter 2, this chapter has thus revealed the complexities involved in exploring the size, growth, location and contributions of the service sector through an analysis of producer services in general and APS in particular. Although such services are rapidly growing and becoming of greater importance as a source of employment, major international and intra-national variations have been identified. Despite some deconcentration of producer services, these services remain clustered in major urban agglomerations and their regions in most advanced economies. Those activities being decentralized, moreover, tend to be back office functions and those serving a local and regional clientele, even though there is some evidence of specialized deconcentration in particular areas.

As with the service sector in general, analyses of the contributions of producer services to economic development have focused upon both their tradability and their relationship to other sectors of the economy. These reveal that producer services play a key role in many economies not only as export activities but also in supporting and aiding the competitiveness of other sectors in the economy. The problem, however, is that although both the export propensity of producer services and the inter-dependencies between producer services and other sectors are recognized, consumer services remain relatively ignored by service researchers. More often than not, they are simply assumed to be locally oriented activities which are dependent upon the other sectors of the economy for their vitality and viability. Here, however, it will be argued that a similar rethinking of the role of consumer services in economic development to that which has occurred in the producer services literature is required. In the next chapter, therefore, this re-evaluation of consumer services is set in motion by examining whether they are indeed evenly distributed parasitic activities dependent on other industries, or whether they are also basic activities which shape, and are shaped by, other sectors of the economy in an inter-dependent manner.

4

CONSUMER SERVICES
A motor for local economic development?

INTRODUCTION

To set the context for the rest of the book, this chapter introduces the reader to the magnitude, character and location of the consumer services sector in Britain and critically evaluates conventional suppositions about its role in local economic development. The aim, in so doing, is to start to transcend the assumption that consumer services are parasitic activities which live off the endeavours of other sectors of the economy and contribute little, if anything, to local economic development.

Consequently, this chapter examines first, the size and character and second, the geography of consumer services in contemporary Britain. To test the assumption of conventional discourse as outlined above, several questions will be posed. Is the growth in consumer service employment reliant upon wealth created by the expansion in other economic sectors? Which are the fastest-growing consumer services? Are they 'luxury' services which, as affluence rises, are increasingly in demand as people with higher disposable incomes devote a greater proportion of their expenditure to a more diverse range of luxury services? And are consumer services over-represented in more affluent areas? The answers to these questions will reveal that orthodox demand-led explanations are wholly insufficient to explain the magnitude, growth and location of this sector. In major part, this is shown to be because these theories erroneously assume that consumer services are dependent activities and solely serve local markets.

Therefore, and so as to move towards explaining the size, growth and geography of consumer services, this chapter retheorizes their role in local economic development. To do this,

40

the following key questions are asked. How, if at all, do consumer services contribute to the revitalization of local economies? Do consumer services generate external income for local economies? Or is this sector merely a dependent activity? And are locally oriented activities less important than externally oriented activities so far as local economic revitalization is concerned? In so doing, this chapter will reveal that consumer services can be motors of local economic development and contribute to local economies not only in their role as external income generators but also as leakage preventers and as activities which support and aid the competitiveness of other sectors of the economy.

THE MAGNITUDE AND CHARACTER OF THE CONSUMER SERVICES SECTOR

As Table 4.1 shows, between 1981 and 1991 in Britain, whilst the number of all employees in employment grew by 2 per cent, consumer service employment rose by 26 per cent. Indeed, 1785000 consumer service jobs were created compared, with just 519000 in producer services (i.e. 3.44 consumer service jobs were established for every producer service job). Consequently, by 1991, more employees worked in consumer services than in the primary and secondary sectors combined and nearly three times as many worked in consumer services as producer services. What is remarkable, however, is that despite its magnitude and growth, this sector, which employs 40.1 per cent of all employees in employment, receives scant regard in the literature on local economic development. Producer services, meanwhile, employing just 10.8 per cent of all employees, and primary and secondary industries which have rapidly declined and together employ only 23.3 per cent of the workforce, remain at the heart of much local economic policy.

The reason for this neglect of consumer services, as stated, is that they are often simply seen as a by-product of the affluence generated elsewhere in the supposedly wealth-creating sectors of the economy. Consequently, as affluence rises, people with higher disposable incomes are asserted to spend a higher proportion of their income on a more diverse range of 'luxury' consumer services (Marshall and Wood 1995). If this demand-led explanation is correct, then the fastest-growing sectors of consumer services should be those which produce these 'luxury' services. However, and as Table 4.1 displays, this is not wholly the case. Although

Table 4.1 Employment change in Britain: by sector, 1981–91[a]

	1981 (000s)	1991 (000s)	% change
Primary and secondary	8338	6704	−19.6
Extractive and manufacturing industries	6652	5026	−24.4
Utilities	1686	1678	−0.5
Service activities	12911	14901	+13.4
Producer services	1819	2336	+28.4
Wholesale distribution/scrap dealing, etc.	876	931	+6.3
Road haulage	194	221	+13.9
Auxiliary services to banking/insurance	88	142	+61.4
Professional/technical services; advertising	199	275	+38.2
Business services	257	580	+125.7
Hiring out machinery, equipment, etc.	47	62	+31.9
Research and development	121	90	−25.6
Trade unions, business and professional associations	37	36	−2.7
Consumer services	6869	8654	+26.0
Retail distribution	2060	2311	+12.2
Restaurants, snack bars, cafés	192	288	+50.0
Public houses and bars	240	339	+41.2
Night clubs and licensed clubs	132	141	+6.8
Hotel trade	226	273	+20.8
Other tourist/short-stay accommodation	36	40	+11.1
Footwear/leather repairs	5	4	−20.0
Repair of other consumer goods	23	18	−21.7
Road passenger transport/urban railways	208	311	+49.5
Hiring out consumer goods	30	33	+10.0
Higher education	198	227	+14.6
School education	1066	1203	+12.8
Driving and flying schools	3	3	0.0
Medical/health/veterinary	1285	1510	+17.5
Welfare/religious/community services	515	896	+74.0
Recreational/cultural	430	512	+19.1
Personal services	120	138	+15.0
Mixed producer/consumer services	2718	2961	+8.9
Canteens	113	117	+3.5
Repair/servicing of motor vehicles	211	171	−18.9
Railways	174	134	−23.0
Other transport	2	1	−50.0
Sea transport	66	32	−51.5
Air transport	70	70	0.0
Support services to transport	100	73	−27.0

Continued

CONSUMER SERVICES

Table 4.1 (continued)

	1981 (000s)	1991 (000s)	% change
Miscellaneous transport services	168	186	+10.7
Postal services	183	206	+12.6
Telecommunications	240	223	−7.1
Banking	368	426	+15.8
Other financial institutions	111	180	+62.2
Insurance	225	271	+20.4
House and estate agents	63	100	+58.7
Legal services	121	186	+53.7
Accountants, auditors, tax experts	104	166	+59.6
Hiring out transport equipment	15	28	+86.7
Owning and dealing in real estate	98	166	+69.4
Education and vocational training	225	304	+35.1
Laundries, dry cleaners, etc.	61	50	−18.0
Public services	1505	1362	−9.5
All employees	21148	21569	+2.0

Source: Census of Employment 1991
Note: a The definition used to categorize service-sector activity into producer, consumer and mixed producer/consumer services is that developed by Marshall *et al.* (1988: Table 2.4)

some of the most rapidly expanding consumer services could be linked to higher disposable incomes (i.e. public houses and bars; restaurants, snack bars and cafés), even if not necessarily to 'luxury' consumption, this must be set against other fast-expanding consumer services (i.e. welfare, religious and community services; scheduled and other road passenger transport/urban railways). These latter industries hardly reflect the increased acquisition of services associated with affluent lifestyles. It appears, therefore, that such a demand-led explanation of consumer service growth is erroneous.

It is not only this conventional explanation for the expansion of consumer services, however, which appears to be inaccurate. So too are alternative explanations. The tendency identified by Gershuny (1978), whereby consumer service jobs are being replaced by manufactured goods as people increasingly engage in 'self-servicing' rather than acquire formal-sector services, is not supported by the data in Table 4.1. All of the fastest-growing consumer service sectors reflect the trend whereby informally produced goods and services (for example community support, home-produced meals, home-brewing) are being replaced by formal-sector

43

services (for example formal welfare support and eating and drinking externally prepared products outside the home).[1] In part, this early argument was a product of its time. Even Gershuny, by the early 1990s, was himself arguing that rather than a shift taking place towards a 'self-service' economy, consumer services would continue to grow because of the 'time-famine' being suffered by people in employment (Gershuny 1992).

Consequently, the recent expansion of consumer services cannot simply be explained in demand-led terms as being due to rising affluence and the increasing demand for luxury services. Throughout this book, therefore, an attempt will be made to move beyond such simplistic reasons and towards a more refined understanding of the growth and nature of consumer services. This is relatively uncharted territory. There have been few attempts to understand this topic. One recent book on the role of services in urban and regional development, for example, devotes just 15 lines to explaining the trends in consumer service demand (see Marshall and Wood 1995: 20). This is by no means an exception to the rule. Neither is it only in the realm of explaining shifts in consumer service demand that major gaps exist in the literature. Similar voids prevail so far as attempts to analyse the geography of consumer services are concerned.

THE GEOGRAPHY OF CONSUMER SERVICES

The common supposition is that consumer services are over-represented in more affluent areas. In other words, the assumption is again that the level of consumer services reflects affluence and the resulting desire for luxury services. As Table 4.2 shows, however, this is far from the reality. Examining their location quotients (LQs), the table reveals instead that it is the poorer peripheral northern regions which are over-represented by consumer service employment whilst such jobs are under-represented in the more affluent southern regions. Yet again, therefore, the commonly accepted wisdom that consumer services are merely a product of affluence is called into question.

Neither is it clear-cut that consumer service employment is over-represented in major urban agglomerations, as suggested by Christaller's central place theory (Christaller 1966). In this theory, which dominates portrayals of the intra-regional distribution of consumer services, such services are seen to group into a hierarchy of urban

Table 4.2 Sectoral distribution of employees in employment in Britain: by region, 1991

	All services		Producer services		Consumer services		Mixed producer/ consumer services	
	No. (000s)	LQ	No. (000s)	LQ	No. (000s)	LQ	No. (000s)	LQ
Greater London	2736	1.18	506	1.44	1265	0.97	717	1.61
South-East	5676	1.11	991	1.27	2908	1.00	1298	1.31
East Anglia	542	0.97	86	1.02	303	0.96	109	1.01
South-West	1246	1.02	162	0.87	732	1.06	230	0.98
West Midlands	1266	0.88	205	0.93	741	0.90	219	0.78
East Midlands	939	0.87	145	0.87	565	0.92	155	0.74
Yorkshire and Humberside	1227	0.93	168	0.83	750	1.01	211	0.83
North-West	1634	0.97	227	0.89	979	1.03	281	0.86
North	719	0.92	78	0.68	456	1.03	105	0.69
Wales	654	0.95	64	0.61	415	1.07	91	0.69
Scotland	1406	0.99	170	0.78	842	1.05	252	0.92

Source: Census of Employment 1991

centres with 'higher-order' goods, which depend on more discretionary spending, locating in larger centres with a higher threshold population, whilst 'low-order' necessities are seen to be offered in smaller local centres. Indeed, this theory for the variable status and spacing of service centres is so accepted that it is one of the few theories to have emerged in geography which shapes, rather than simply reflects, reality in that contemporary urban and regional planning policy is based on implementing such a view of the world.

As Marshall and Wood (1995) assert, however, a number of trends have worked against such a simplistic hierarchical description of consumer service location: growing consumer mobility; increased consumer differentiation (for example by age, gender, social grouping); the locational strategies of increasingly dominant consumer service organizations; changing household work patterns; counter-urbanization; the uneven distribution of the growth of incomes and spending power; demographic trends; changing regulatory frameworks; and the development of centres of specialized consumption. The result is that consumer service employment is under-represented in Greater London, the highest-order urban centre in the UK, compared with the rest of the country (see

Table 4.2). Similar tendencies are to be found at the next level down the urban hierarchy. As chapters 11 and 12 display, despite being widely acknowledged as the highest-order regional centre for Yorkshire and Humberside, Leeds is under-represented in consumer service employment whilst Sheffield, a lower-order sub-regional centre in the same region, is over-represented.[2]

From this, it is evident that a new geography of consumer services must be forged. To do this, several important questions will need to be addressed. How are the various sub-sectors of consumer services spatially distributed? Is it the case, as in manufacturing and producer services, that there is increasing inter-regional and inter-urban specialization in consumer services? Is there an evolving corporate spatial division of consumer services by function? For example, are large metropolitan regions such as London harbouring this sector's headquarters and related high-level consumer services, whilst other areas generally house lower-order consumer services? Does this spatial division of control functions in consumer services reinforce, or mitigate, the spatial inequalities produced by the other sectors of the economy? Indeed, what are the impacts on uneven development of market concentration in consumer services? How is this changing over time? These are all key questions which will be discussed in both the examination of specific consumer service industries in Part II and the locality studies in Part III of this book so that a new understanding of the spatial divisions of consumer services can emerge.

However, it is not only a new geography of consumer services which is required. The contribution of consumer services to local economic development also demands major reconsideration since conventional perceptions are similarly grounded in the belief that such services are reliant upon the wealth created in other sectors of the economy. Consequently, in the next section, an attempt is made to move beyond this dependent sector characterization and to start to retheorize the role of consumer services in local economic development.

RECASTING THE ROLE OF CONSUMER SERVICES IN LOCAL ECONOMIC DEVELOPMENT

How, if at all, do consumer services contribute to the revitalization of local economies? Do consumer services generate external income for localities? Or is this sector merely a dependent activity?

And are locally oriented activities less important than externally oriented activities so far as local economic revitalization is concerned?

Consumer services: dependent activity or engine of growth?

Despite the abundance of literature which displays that producer services export their products, and their consequent redefinition as basic-sector activities, consumer services have been subject to little or no equivalent investigation. Instead, they frequently remain relegated to the subordinate status of being dependent activities. Lundmark (1995: 137) epitomizes this view of consumer services when he asserts, 'Many producer services contribute directly to the economic development of regions, in the sense that they are producing not only for a local or regional market (in contrast to most consumer services).' Yet, he makes no attempt to test this assertion so far as consumer services are concerned.

In part, such a characterization of consumer services is due to the narrow definition of the basic sector adopted in many studies. Because basic activities have been perceived as only those which export their products as opposed to generating external income by importing consumers, the result has been that although producer services have been incorporated, consumer services have not. However, given that consumer service industries which induce external income into an area fulfil the same function as export industries, in that both generate external income so as to allow the economy to grow, this conceptualization of the basic sector has been increasingly questioned (for example Farness 1989, Hefner 1990, Marshall and Wood 1995, Williams 1996a). Consequently, two types of 'basic'-sector activity have been identified. First, there are export industries, which include primary, manufacturing and an array of producer services which sell their products to customers outside the area, and second, there are those industries which draw consumers into a locality in order to spend their money. Since both fulfil the same function of earning income for a locality, both have started to be accepted as basic activities.

Despite this, much local economic policy has continued to focus on developing and attracting export-oriented activity (Persky et al. 1993). 'Basic consumer services' which attract consumers into the area in order to spend their money have often received much less attention. Even when tourism, the basic consumer service most

47

commonly promoted by development agencies, is fostered, it is still treated as somehow different from the supposedly mainstream economic development activities of local authorities, as will be revealed in chapter 5. Nevertheless, at least it receives some consideration in the context of economic development. Local economic policy has until recently paid much less attention to many other consumer services which also perform a 'basic'-sector function, such as medical services (Gross 1995, Vaughan *et al.* 1994), sports facilities and events (see chapter 6), higher education establishments (chapter 7), retailing (chapter 8) and the cultural industries (chapter 9).

The outcome is that most development agencies have pursued inward investment in the shape of manufacturing and many chase producer services, but fewer have endeavoured to lure the full range of basic consumer services. Take, for instance, the retail sector, discussed in chapter 8. Few local authorities recognize retail revitalization as being an issue in their economic development strategies, some failing even to mention it when their authority is actively raising its town centre up the retail hierarchy so as to draw in more external income. Such a blindness to the value of basic consumer services is not confined to retail developments. It is also clearly depicted in the provision of sports and leisure facilities. To witness the way in which such provision is regarded as a 'parasitic' rather than a basic activity, one need look no further than the commotion caused by Sheffield City Council's decision to host the World Student Games (see chapter 11) or, more generally, by the fact that the proposal for a sports or leisure centre in a locality is nearly always perceived as a cost to the local authority rather than an economic investment. Hence, although producer services have been accepted as basic activities in much local economic policy, the same cannot be declared about the full range of 'basic consumer services', despite their income-inducing abilities.

The evidence in both the sector-specific studies and the locality studies throughout this book will reveal that this is a mistake. Here is just one example. Using a series of commodity-by-industry input/output models for the Canadian provinces, Stabler and Howe (1993) compare the exportability of producer services and personal and other miscellaneous services. They reveal that this consumer service not only sells its product inter-regionally and internationally but, in terms of gross sales, induced more income into Canadian provinces from outside than producer services in

CONSUMER SERVICES

1984 ($5367 million compared with $4491 million). As Stabler and Howe (1993: 41) conclude, the consequence is that we should not discuss any services as 'non-tradeables': 'there are very few services that are not traded both interprovincially and internationally'.

Basic consumer services, therefore, are external income generators in the same way as manufacturing and producer services. As Marshall and Wood (1995: 169) assert, 'while the local economic bases of manufacturing, finance and businesses services are important influences for any urban economy, the evidence of the 1980s shows there is considerable scope for independently supported consumer service growth'. Reflecting this, a number of localities have started to shift their economic development strategies from importing producers to importing consumers. The problem, however, is that this reinterpretation of economic base theory still does not go far enough. As will now be shown, although it transcends the basic-/dependent-sector dichotomy which relegates consumer services to the dependent-sector category, it fails to question the assumption that a hierarchy of sectors can be discerned based upon their ability to earn external income.

Rethinking the contribution of externally and locally oriented activity

To grasp the full relevance of consumer services to local economic development, one has to shift beyond an appraisal of their basic-sector characteristics alone since this presupposes that economic growth is a direct result of the level of external income generation. However, research has revealed that economic growth is not so strongly correlated with income-inducing activity as many previously assumed (Giaratani and McNelis 1980, McNulty 1977, Mandelbaum and Chicoine 1986). As will be shown in subsequent chapters, this is because what is needed for an economy to grow is not a rise in external income alone but, rather, an increase in *net income*. Net income, to reiterate, is determined by total external income, times a multiplier (which is larger the more self-reliant the economy), minus total external spending (Williams 1994a). The growth of a local economy is thus dependent on not only attracting external income but also preventing the leakage of money out of the area (Persky *et al.* 1993). As Power (1988) asserts, the only reason to export is to pay for imports. If imported goods and services can be produced locally, then net income will increase without a rise in exports. The variety of economic activity then increases, as does the interconnectedness of the local economy.

At present, however, the positive role that locally oriented activities can play in economic growth as leakage preventers is seldom considered. As Harrington *et al.* (1991: 91) assert, 'the issue of service imports has not often been addressed by advocates of service development as a form of regional policy, since such writers have been more concerned with the export potential of certain services'. Instead, most development agencies pursue export-led development policies by cultivating the basic sector of their economy with little regard for the extent to which seepage of income is taking place (Persky *et al.* 1993). The consequence is that many local economies leak very badly. Indeed, so rarely is such a notion considered important that data on the extent to which local economies leak are very rare. One exception is the work of Polese (1982: 158), who finds that over half of regional service demand in a rural area of Quebec is satisfied by imports. Consumer services, therefore, perform two functions in local economies. First, they can act as 'basic activities', drawing income into the economy from outside, and second, they can operate as 'leakage preventers' serving the local market and thus retaining and circulating money within the locality. These functions are equally important since both are required to increase net income. Indeed, and as will be shown in the forthcoming chapters, a consumer service can frequently perform a basic-sector function of inducing income into a locality whilst, at the same time, retaining income within the area which would otherwise drain out.

Hence, and contrary to popular interpretations of economic base theory, locally oriented activities can make a positive contribution to economic development. When such activities substitute for imported services, they lead to economic growth. As Gillis (1987) notes, import substitution in services is as important a goal as export promotion. Take, for example, the provision of medical services. Whilst a number of studies undertaken in North America (for example Gross 1995, Vaughan *et al.* 1994) have focused upon the export-generating potential of such medical services, Gillis (1987) shows that a new medical centre also contributes to economic development if it provides services previously unavailable in the locality and thus retains income.

The implication, therefore, is that it is a mistake to examine the ability of an industry to generate external income as the measuring rod of its contribution to local economic development. Instead, equal emphasis must be given to both leakage prevention and external income generation. Put another way, creating denser inter-linkages and synergies between all sectors of the economy so as to meet local demand is just as important as generating external income.

CONCLUSIONS

From the above analysis, it is clear that the relationship between consumer services and economic development is in need of a major rethinking. As shown, neither the magnitude, growth nor location of consumer services can be adequately explained by simplistic demand-led explanations. This is because these assume consumer services to be dependent activities and purely serving local markets. Here, however, and in a bid to move towards an understanding of the size, growth and location of consumer services, the role of this sector in local economic development has been reconceptualized. It has been suggested that consumer services not only function as basic-sector activities but also contribute to economic development in their role as leakage preventers. Both are propounded to be equally important for local economic development.

Based on this rethinking, the intention in the rest of this book is to start to unravel some of the complexities involved in recasting the role of consumer services in local economic development. In order to do this, attention now turns in Part II to an examination of specific sub-sectors of consumer services, followed in Part III by an examination of the role of consumer services in local economic development in a variety of localities.

Part II

THE ROLE OF SPECIFIC CONSUMER SERVICE INDUSTRIES IN REGENERATING BRITISH LOCALITIES

5

TOURISM AND ECONOMIC REGENERATION

INTRODUCTION

Compared with other consumer services, tourism is widely accepted as a tool for revitalizing local economies. Nevertheless, it is not always treated in the same manner as other economic sectors. In local authorities, for instance, tourism strategies are usually separate from, rather than a part of, economic development strategies, and responsibility for this function is frequently distinct from economic development (Williams 1996c). In academe, similarly, tourism is often cordoned off from the study of economic regeneration. Economic geographers, for example, concentrate overwhelmingly on manufacturing and producer services. Few deem tourism a subject worthy of their attention (Ioannides 1995). This is reflected in, and reinforced by, tourism research mostly being published in journals specifically dedicated to the subject (for example *Annals of Tourism Research, Tourism Management* and *Journal of Sustainable Tourism*). Relatively little research finds its way into mainstream economic development debates and fora. As such, tourism remains an island onto whose shores few mainstream economic development commentators drift.

Consequently, the aim of this chapter is to provide an overview of the role of tourism in local economic development. First, the nature and extent of tourism will be outlined, followed by an analysis of the magnitude of, as well as the problems and issues involved in assessing, its local economic impacts. This will reveal that tourism is a major and growing sector of the economy which contributes to local economic development not only as an external income generator but also frequently as a lead sector which seeks to improve the image of an area in order to attract other investment. However, this chapter concludes by arguing that its impacts could be improved with several changes in how tourism is conceptualized. These alterations are outlined.

Here, and following Law (1990), tourism is defined as the attraction of visitors, both staying and day, drawn from outside the normal travel to work area, whether for pleasure (attractions) or business (conferences, exhibitions). As such, it includes not only holiday travel but also business trips and conventions. Although visiting business people are sometimes excluded from studies of tourism since their visit is work-related, they are undoubtedly visitors who have been attracted to the locality and are spending external income, just like any other tourist.

THE CHARACTER AND MAGNITUDE OF THE TOURIST INDUSTRY

To commence, the nature of the tourist industry will be analysed; this will be followed by an examination of the extent and geography of tourism. The intention is to analyse the potential of this consumer service industry as an economic development tool.

The character of the tourist industry

Tourism is an ever more heterogeneous industry. There is an increasing diversity in: the types of place tourists visit (for example seaside resort, rural area, urban industrial heritage, capital city); the main purpose of their visits (for example sun-seeking, adventure, business, culture, sport); the forms of accommodation used (for example camping, five-star hotel, bed and breakfast, self-catering, staying with friends or relatives); the degree of organization sought (for example from package tours with pre-arranged visits to independent travelling); duration of stay (for example a day visit, short break or long stay); and the composition of tourist groups (for example alone or with family, a hobby club, business colleagues or similar-aged people). Indeed, the place, purpose, accommodation, length of stay, degree of organization and composition of the tourist group can be combined in a multitude of different ways to produce an industry which is composed of a diverse mix of provision.

In part, this diversification is due to demand-led factors such as changing tourist preferences and greater differentiation in the desires of different groups. However, it is also supply-led with the range of opportunities open to tourists having expanded as localities have sought to mitigate their economic problems by actively marketing and constructing a greater diversity of tourism products (Bramwell and Rawding 1994, 1996). Deindustrialization in old industrial urban areas, for example, has led to a raft of new products including urban tourism (for example van den Berg *et al.* 1995, Bramwell 1993, Law 1990, 1992, Page 1995, Roche 1992), heritage tourism (Light and Prentice 1994) and cultural and sports tourism (see chapters 7 and 8), whilst agricultural problems in rural areas have resulted in the rise of rural, green and sustainable tourism (for example Bramwell and Lane 1994, Lane 1990, Hunter and Green 1994).

Neither is demand for these different forms static. As Urry (1995) asserts, there is a complex and ever-changing hierarchy of the objects desired by tourists resulting from this interplay between both changing class, gender and generational tastes and competing capitalist and state interests involved in the provision of such objects. Indeed, so great has been this interest in feeding the 'tourist gaze' (Urry 1990) that for some, we are at the 'end of tourism' because the purchase of images has become so widespread that it is now no longer confined to specific tourist practices (Lash and Urry 1994). A major industry has emerged which cuts across other sectors of the economy. A shopping trip, for example, is becoming an opportunity to engage in consumer myths and fantasy similar to traditional tourist practices (Goss 1993, Shields 1991). Seen in this light, tourism has become inextricably interconnected with all sectors of the economy and with everyday life itself.

Identifying the tourist industry is thus a difficult task. It is composed of numerous industries which vary temporally and spatially in the extent to which they rely on tourism spending. Indeed, just as there are few producer services which are solely dependent on intermediate demand (see chapter 1), there are few industries which can be described as entirely reliant on tourist expenditure, perhaps with such notable exceptions as the hotel sector and travel agencies. Restaurants, for example, whilst generating a proportion of their income from tourists, cannot be characterized as a pure tourist industry. Equally, there are a number of supposedly 'non-tourist' industries such as banks, supermarkets, insurance companies and aircraft manufacturers which receive a proportion of their income from tourists but are not considered tourist industries. In reality, therefore, tourism is a collection of industries and public agencies that directly or indirectly, and to differing degrees, enable the production of tourism experiences.

Nevertheless, amongst those industries which earn a higher than average proportion of their income from tourists, some key trends can be observed. Examining airlines, travel operators and the hotel industry, Ioannides (1995) identifies increasing market concentration through both horizontal and vertical integration and a high level of product standardization. In the hotel sector, for instance, the proportion of hotels on a global level belonging to multi-site establishments increased from 19.1 per cent to 21.2 per cent between 1980 and 1986 (World Tourist Organization 1988).[1] In the airline industry, meanwhile, similar oligopolistic structures are evident, especially in the US and Europe. By December 1991, the five largest US carriers accounted for 77.8 per cent of total passenger miles flown in the US compared with only 55.4 per cent in 1985. The three largest (American, United and Delta) accounted for 55 per cent of all passenger miles flown and 66.4 per cent of all revenues (Ioannides 1995).[2] In tour operations, similarly, the share of inclusive tour holidays sold by the top five UK operators (out of a total of about 680) rose from 50 per cent to 70 per cent between 1984 and 1988. By 1989, it had reached 77.5 per cent (O'Brien 1990). Such horizontal integration, however, is not the sole means by which market concentration is occurring. There is also a process of vertical integration. Airlines are expanding into tour operating, tour operators acquiring airlines and travel agencies investing in the accommodation sector (Ioannides 1995, Urry 1990). Hence, major corporations are now able to control tourism flows (Britton 1991) and can play a pivotal role in shaping the fortunes of individual localities.

For some, this vertical and horizontal integration is resulting in a high degree of standardization in tourism provision (Ioannides 1995). However, although this may be true of the experience of hotels, airlines and tour organization it is not necessarily the case so far as the overall tourist experience is concerned. Global processes pass through the lenses of local history and culture which transform their manifestation, leading to diversified products for the 'tourist gaze' (Urry 1990). Indeed, some have argued that far from the creation of homogeneous places, the result is more an 'individualistic spatiality' (Soane 1993). Whether or not standardization is occurring in tourism products, there is little doubt that the nature of tourism is undergoing rapid change as forms of tourism diversify and its magnitude expands.

The magnitude of the tourist industry

The growth of the tourism industry is a late twentieth-century phenomenon (Murphy 1985). In 1960, there were just 60 million international tourist arrivals worldwide, but 280 million by 1982 and 429 million by 1991. The result is that tourism is now a major industry. International tourists alone spend $209 billion a year and generate at least 60 million jobs (Urry 1995). This expansion, however, cannot be explained in purely demand-led terms, as is commonly assumed in the case of such consumer services (see chapter 4). Although rising general affluence is an important determinant, this growth is also due to supply-side factors such as technological advances in transport and developments in the tourist industry itself.

Neither can this consumer service industry be argued to be evenly spread throughout the world. Just as the vast majority of service-sector trade is accounted for by the OECD states (see chapter 2), similar patterns can be discerned in the tourist industry. Some 90 per cent of international travel is accounted for by OECD nation residents (Jefferson 1995), and Europe specifically is both the largest generator of, and the major destination for, international tourists. Fifteen of the top twenty nations generating tourists are European (Jefferson 1995) and 64 per cent of all international tourist arrivals worldwide in 1991 came to Europe, which possesses eight of the top ten tourist destinations in terms of tourist receipts (Urry 1995).[3] The tourist industry, therefore, is heavily concentrated in Europe, generating some 7.5 million jobs in the EU alone (Williams and Shaw 1990).

Tourism is also a significant segment of the export base of these economies. In Britain, for instance, the annual turnover from tourism was some £19 billion in 1989, with 1.37 million people employed in the industry, or 6.2 per cent of all employees (see Table 5.1). Overseas visitors spent £6850 million in the UK in 1989, which represents 28 per cent of the country's service export earnings (more than the City of London) or 4.6 per cent of the country's total export earnings (House of Commons Employment Committee 1990). It is an industry, moreover, which is continuing

Table 5.1 British tourism statistics

	Jobs in tourism (m)	% of total jobs	Visits by overseas visitors (m)	Earnings from overseas visitors (£m)	Visits by domestic visitors (m)	Spending by domestic visitors (£m)
1983	1.14	5.6	12.5	4003	131	5350
1984	1.20	5.8	13.6	4614	140	5975
1985	1.27	6.1	14.4	5442	126	6325
1986	1.27	6.1	13.9	5553	128	7125
1987	1.28	6.1	15.6	6260	132	6775
1988	1.32	6.1	15.8	6193	130	7850
1989	1.37	6.2	17.2	6850	110	10875

Source: House of Commons Employment Committee (1990: Table 1)

to grow both absolutely and relatively. Overseas earnings from tourists rose by 71 per cent between 1983 and 1989, whilst the number of employees in tourism rose by 20 per cent, compared with 5 per cent for the British workforce as a whole (House of Commons Employment Committee 1990).

More recent figures show that this growth in tourism is continuing during the 1990s. In 1994, Britain received more than 21 million visits from overseas tourists (8 per cent more than in 1993), who spent almost £10000 million, and some 1.5 million people were employed in the travel and tourism industry (Foreign and Commonwealth Office 1995). In 1995, moreover, Britain set new records registering 23.6 million visits, 12 per cent more than in 1994, and spending was up by 17 per cent to £11700 million. The result is that the foreign income from tourism in 1995 was greater than the total exports of either North Sea oil or the financial services industry (Foreign and Commonwealth Office 1996).[4]

If the tourist industry represents a large component of the export base in many developed nations, the same applies to local and regional economies. Again, however, such activity is unevenly distributed. To show this, attention turns towards an analysis of the spatial manifestations of tourism in contemporary Britain.

The geography of tourism in Britain

Whether the tourist industry reinforces or mitigates the inequalities between the core and periphery has been the subject of much discussion (see Shaw and Williams 1994). The traditional view represented by Peters (1969: 11) is that 'tourism, by its nature, tends to distribute development away from the industrial centres towards those regions in a country which have not been developed'. Christaller (1963), similarly, argues that tourism can be a means of achieving economic development in peripheral regions since the flow of rich tourists from the 'metropolitan' core should inject foreign exchange and generate jobs.

However, this static view of tourism is highly simplistic. Tourist products do not exist in any absolute sense but, instead, are the result of investments and marketing. They are the outcome of both demand- and supply-led factors. Tourism, that is, is a dynamic industry whose products are continually changing in reaction to both shifting consumer preferences and wider structural processes (see, for example, Townsend 1992). To understand whether tourism reinforces or mitigates the spatial inequalities produced by other sectors, and how this is altering over time, Williams and Shaw (1990) thus examine the different forms of tourism which are emerging and their distinctive spatial implications (see Table 5.2).

This reveals that different forms of tourism have varying spatial impacts. Take, for example, international conference and exhibition tourism. As with the geography of producer services discussed in chapter 3, despite some limited concentrated decentralization to

Table 5.2 Typology of the regional implications of tourism developments

Type of tourism	Locational bias		Selective regions			
	Core	Periphery	Capital	Urban	Rural	Coastal
International conferences and exhibitions	*		*			
International cultural	*		*			
Industrial heritage		*		*	?	
Conference	*	*	*	*		*
Exhibition	*	*	*	*		?
Business	*			*		
'Events'	*	*	*	*	*	*
Mass summer		*			*	*
Mass winter		*			*	
Short break (summer)	*				*	*
Short break (winter)	*		*	*		

Source: Williams and Shaw (1990: Table 2)

61

second-tier cities (for example Birmingham, Leeds, Glasgow, Edinburgh, Brighton, Bournemouth), London remains the major location for such activity (Hughes 1988, Law 1992). Similarly, international cultural tourism remains centred on London (see chapter 14), despite the attempts by second-tier cities such as Glasgow to capture a greater share of this market (see chapter 10). These growth sectors of tourism, therefore, reinforce spatial disparities. Nevertheless, other forms of tourism are more periphery-oriented, such as heritage tourism in old industrial areas, rural tourism and mass winter breaks. Each form of tourism, therefore, has its own distinctive spatial pattern and different implications in terms of its contributions to core/periphery patterns of development. These, furthermore, shift over time. Williams and Shaw's (1990) typology, although tentative, is thus a useful heuristic device for exploring the spatial implications of different forms of tourism.

Overall, however, there is little doubt that tourism presently reinforces many of the wider spatial inequalities. As will be shown in chapter 14, London lies at the heart of the British tourist industry, capturing 53.8 per cent of all overseas tourist expenditure and 54.5 per cent of overseas visits and harbouring the vast majority of headquarters of tourist-related corporations in the hotel, airline and travel operator industries. This concentration, moreover, is increasing over time. Nevertheless, comparatively little work has been conducted on delineating the spatial disparities of either tourism as a whole or these different forms of tourism, and much more work is required before the complexities can be understood fully. As will now be shown, this is because much tourism research has focused upon its economic impacts on individual localities.

THE LOCAL ECONOMIC IMPACTS OF TOURISM

During the last two decades, there have been a multitude of local economic impact studies of tourism. These measure three types of economic impact: direct impacts, which are the first-round impacts of the additional spending which visitors bring to a locality; indirect impacts, which include the ripple effects of additional rounds of recirculating the initial external visitors' spending and the spending induced from the wages and salaries paid to workers employed to cater for these tourists; and intangible impacts, which cover the impacts of tourism on such matters as the image of the locality in the eyes of potential investors and locals. Here, the problems and issues involved in assessing each of these impacts will be considered in turn as well as their size. To do this, the considerable literature on the local economic impacts of tourism will be drawn upon.

Direct impacts

The principal reason for advocating tourism as a tool for economic revitalization is that it directly generates external income by inducing visitors into a locality who subsequently spend money. Thus, the simplest measure of the benefits of tourism is the amount of external income it produces for a locality. This can be significant. On Merseyside, for example, Vaughan (1990) calculates that in 1985, visitors spent a total of £223 million, or £95 million when business tourists are excluded. In London, meanwhile, British and foreign tourists spent £6000 million in 1995 (Foreign and Commonwealth Office 1995), whilst in New York, the 19 million overnight tourists spend $1.3 billion annually, which generates more than 117100 jobs in the New York metropolitan region (Port Authority of New York and New Jersey 1993). Tourism, therefore, is a major industry whose receipts are often far in excess of the dominant manufacturing and producer service sectors in their areas (see chapters 11 and 14).

Determining direct expenditure, however, is problematic, especially when analysing a particular tourism event or project. On the one hand, there is the problem of determining whether all the initial expenditure can be attributed to the tourism event or project under consideration. Some visitors, for example, may be 'time switchers', having changed their visit to the locality in order to fit in with the particular tourism event being analysed. Their spending cannot therefore necessarily be attributed to the event, unless it has increased their stay and spending in the locality. On the other hand, there is the problem of deciding whether spending by local people should be ignored since it is merely an internal transfer (Hughes 1994). Although it is common to exclude expenditure by locals, studies such as that by Johnson and Moore (1993) of white-water recreation in Oregon find that ignoring all local users' expenditures because they represent the recirculation of money is erroneous. Many of the local users would have gone outside the locality were the facilities not available, and their expenditures would have been lost to the area. Even if these problems with estimating direct expenditure are resolved, further problems confront the researcher.

Multiplier effects

Total visitor spending alone does not provide a complete picture of tourism's economic benefits. As this initial round of spending seeps through the economy, it provides direct revenue for many businesses so that the general output of the area rises (assuming that sufficient resources are available), employment opportunities increase and personal incomes grow. These are the 'indirect' effects. As wages and salaries within the economy increase, moreover, so local consumption rises, and this provides a further impetus to economic activity, generating additional turnover. These are the 'induced' effects (Archer 1977).

Although such multiplier effects are widely known and routinely used when measuring tourism's economic impact, there has been a good deal of critical evaluation of the problems involved in generating them (Archer 1982, 1984, Butler 1993, Fletcher 1989, Fletcher and Archer 1991, Fletcher and Snee 1989, Hughes 1994, Johnson and Thomas 1990, Mathieson and Wall 1982, Vaughan 1990, Vaughan and Wilkes 1986). Here, the most common problems encountered are briefly reviewed in order that the reader can critically appreciate the issues involved in conducting such analyses.

As outlined by Fletcher and Snee (1989) for instance, the first issue is to decide whether to examine how tourist expenditure changes either output, sales/transactions, income, employment, government revenue or imports. Of these six multipliers, it is widely agreed that employment multipliers are the least reliable. An employment multiplier measures the direct, indirect and induced effect of an extra unit of visitor spending on employment in the locality, showing how many full-time equivalent jobs are created. On the assumption that all existing employees are fully utilized, employment is commonly estimated by converting expenditure into units of labour, with the further assumption that increased employment will result from an increase in expenditure/output in each sector and that it will increase in a linear form. However, in practice, increases in output may be met through increased utilization of existing capacity, such as by asking people to work overtime, and/or there may be a less than proportionate increase in employment. This is particularly the case with short-term events (Burgan and Mules 1992). Neither can it be assumed, as is commonly the case, that unemployed labour and mobile and substitutable resources exist to feed this increased demand (Hughes 1994).

The income multiplier, in contrast, measures the effect of an extra unit of visitor spending on local household incomes and is widely used 'because it reflects most clearly the economic impact on host community residents' (Murphy 1985: 92). A sales or transaction multiplier, meanwhile, measures the effect of an extra unit of visitor spending on business turnover. Sales or transaction multipliers, however, produce higher overall impact assessments than income multipliers since household income is only one of five potential recipients of the money spent on a sales transaction; the others are local industries, local government, non-local governments and non-local industries and individuals (Crompton 1995).

Once the multiplier has been chosen, the next problem is to calculate the size of the multiplier effect. The wider the area is defined, the higher will be the local multiplier because of the smaller leakages, but the lower will be the relative impact on overall economic activity. The extent of leakages, moreover, varies not only according to the spatial scale considered but also according to the structure of the local economy. In an entirely self-sufficient economy, the leakage will be zero. In a small community fully integrated into the wider economy, however, the local economic effects of tourism expenditure will quickly diminish as supplies of goods and services are purchased externally and profits repatriated out of the area (Archer 1977, Hohl and Tisdell 1995, Hughes 1982, Vaughan 1990). Reflecting this, Murphy (1985) finds that local multipliers vary between 0.32 and 1.67 (where the value represents the total extra income generated for each unit of tourist expenditures), depending upon how the 'local' area is defined and the structure of the local economy.

Some assert that tourism compares well with other sectors in terms of its multiplier effects on account of the extent of local purchasing (Duffield and Vaughan 1981). This, however, depends upon the form of tourism considered and the structure of the tourist industry in the locality. For example, 'enclave tourism', where tourists are encouraged to remain within a hotel-style complex and food, drink and facilities are included in the original price paid (for example Butlin's, Pontin's and more recently Center Parcs), has very low multiplier effects (Freitag 1994). In areas dominated by externally owned tourist trans-national corporations (TNCs), moreover, multiplier effects will be lower since they source externally to a greater extent than locally owned tourist businesses. Nonetheless, if there have been difficulties in deciding how to assess the multiplier effects of tourism, then the issue of the intangible effects has been even more troublesome.

Intangible impacts

Although the direct and multiplier effects are the quantifiable economic impacts of tourism and have been the focus of much of the work, there are other economic implications which are much less tangible. These include the impacts on the image of the area and whether tourism helps develop positive attitudes towards the locality both for the people who live and work there and for those considering investing or relocating.

Although such intangible impacts are difficult to calculate, they cannot be ignored. It is frequently asserted, for example, that tourism changes the image of a locality, which attracts the attention of those making business location decisions, meaning that the economy will then grow in areas completely unrelated to tourism (Okner 1974). Similarly, although civic boosterism resulting from tourism is difficult either to measure or to value, it cannot be overlooked. Tourism and tourist events, after all, may aid the development of a sense of identity, loyalty and allegiance to an area (Maguire 1995). These intangible impacts will be evaluated in greater depth in the context of tourist-related sports events and facilities in chapter 6. For the moment, it will suffice to say that measuring the intangible impacts of tourism development is a problematic issue and one that has more often than not been ignored by tourism researchers seeking to specify the contribution of this industry to local economic development.

Negative impacts

If the intangible impacts of tourism have been given relatively scant attention by researchers, so too have the negative impacts. Since the 1970s, tourism research has focused upon its positive contributions to local economic development using the above economic

66

impact models to justify to public-sector agencies the validity of investing in tourism development.[5] Although such an approach remains common today, since the 1980s, studies have also begun to explore the negative economic, socio-cultural and environmental impacts of tourism (for example Britton 1978, Ioannides 1995, Mathieson and Wall 1982, Zurick 1992).

In the realm of negative economic impacts, four issues have been highlighted. First, tourism has been conceptualized as a form of dependent development, especially in a Third World context, because of increasing market concentration and greater external ownership and control of the tourism production system (Britton 1991, Ioannides 1995). Second, there are its negative impacts on local labour markets. The perception is that tourism exhibits low productivity levels and heavily depends on low-paid, low-skilled, part-time, occasional and female-oriented labour. The development of tourism, therefore, is seen to lead to the growth of not only peripheral employment but also informal economic activity (Ioannides 1995, Leontidou 1988, Williams and Thomas 1996). Roche (1992), however, argues that this largely reflects the immaturity of an industry subject to seasonality as well as the mono-culture of resort areas. This, he asserts, will diminish over time. The seasonality problem will be increasingly tackled by investment in all-weather tourist attractions (for example domed multi-purpose leisure centres), the development of off season short-breaks and counter-seasonal programming of events. Meanwhile, the labour problems typically associated with isolated mono-cultural resort areas, he asserts, are inapplicable to the emerging tourist localities, such as cosmopolitan urban areas, where facilities are typically designed for both the local resident population and visitors.

A third economic problem of tourism concerns the way in which under market and marketing pressure, there has been increasing competition for tourists. This is assumed to be negative because tourism demand is perceived as fixed. However, others argue that on account of high elasticity of demand in this sector, competition is not so negative as some assume (Williams and Shaw 1990). Indeed, the rapid growth of tourism identified above supports such an argument. A fourth and final economic difficulty is asserted to be the insufficient regard paid to the opportunity costs of investment in tourism by state agencies: what could have been produced if the money had been invested elsewhere? As will be

demonstrated in chapters 10 and 11, however, for many deindustrializing urban areas such as Sheffield and Glasgow, few other
opportunities were available to them.

Besides such economic issues, there has also been closer scrutiny
of the negative socio-cultural impacts (for example Poirier 1995),
including whether tourism always benefits local residents. For
example, the facilities which tourists appreciate may not be the
same as those demanded by locals, may not appeal to locals, or may
be out of their financial reach (see chapter 9). Creating improved
environments may also force locals out of particular areas or
completely change the character of the place (Law 1990, 1992),
as has been witnessed in central Glasgow (see chapter 10). There
has also been an increasing emphasis placed on the negative
environmental impacts of tourism (Bramwell and Lane 1994, Cater
1995, Coccossi and Nijkamp 1995, Hunter and Green 1994). Here,
analysts reiterate in the context of tourism many of the environmental management arguments applied more generally. Cater
(1995), for example, explores the environmental contradictions in
sustainable tourism, and Lane (1990) examines what constitutes
sustainable tourism in rural areas of Britain, whilst Hohl and
Tisdell (1995) call for a 'user pays' principle in tourism.[6]

To conclude, therefore, this review of the local economic impact
studies of tourism reveals that although tourism mostly has positive economic impacts on local economic development, assessing
the exact impacts of tourism is an imprecise science based on many
value judgements at each stage of the analysis. To show in more
detail how these infiltrate such investigations, the next section
employs a case study.

The impacts of Association of American Geographers (AAG) conferences

Randall and Warf (1995) analyse the economic impacts of AAG
annual meetings on the communities and states in which they have
been held. They find that, on average, AAG conferences have
positive economic impacts. The aggregate expenditures of $1.55
million raise state (provincial) output by $2.86 million, create 90.9
full-time equivalent jobs and generate $717000 in personal income
(see Table 5.3). To arrive at this, they use AAG attendance figures
from the AAG office, disaggregated by students and non-students,
and extrapolate the average durations of stay and aggregate expen-

Table 5.3 Economic impacts of AAG conferences, 1983–94

City	Year	Total expenditure ($m)	Total state output ($m)	Job impacts	Personal income ($000s)
San Francisco	1994	1.97	3.64	115.6	911.9
Atlanta	1993	1.79	3.32	105.1	829.4
San Diego	1992	1.60	3.02	92.4	751.0
Miami	1991	1.54	2.84	90.0	710.5
Toronto	1990	2.53	4.67	148.0	1167.8
Baltimore	1989	1.91	3.53	112.1	884.3
Phoenix	1988	1.20	2.21	70.1	553.5
Portland	1987	1.04	1.92	60.9	480.6
Minneapolis	1986	1.27	2.35	74.5	588.1
Detroit	1985	1.24	2.28	72.4	571.5
Washington, DC	1984	1.63	2.99	97.3	739.9
Denver	1983	0.89	1.64	52.1	411.1
Average		1.55	2.86	90.9	716.6

Source: Derived from Randall and Warf (1995: Tables 1–4)

ditures from an e-mail survey of 24 students and 61 non-students attending the San Francisco conference. They then use the estimates of various local tourist boards and economic development authorities to estimate the sectoral composition of these expenditures and employ input/output analysis to estimate the impacts on each state.

The assumptions and values employed to reach such an estimate of the impacts of AAG conferences, however, are manifold. First, the survey used a relatively small sample derived from just one conference. No indication is given of the representativeness of this sample or of the reliability of the data collected. Second, there is no consideration of the applicability of these data to other conferences, for example the fact that attenders at the San Francisco conference might have stayed longer and spent more than at other AAG conferences because of the attractiveness of the location. Third, all expenditure is attributed to the AAG conference. There is no indication of what proportion of conference attenders were 'time switchers' who were going to visit each city anyway but made their visits coincide with the conference. Fourth, behind the calculation of the sectoral distribution of expenditures on the basis of estimates from local tourism and economic development officials in each state, there is an assumption that AAG attenders distribute their expenditures in the same way as tourists in general. This is by no means certain. Fifth, when input/output estimates are used to calculate the employment multipliers, the assumption is that the rise in output will result in new jobs. In practice, however, increases in output may be met through increased utilization of existing capacity, as has been discussed earlier. Sixth, the fact that the area is defined as the state in which the conference was held has a tendency to result in larger multiplier effects since higher proportions of the goods and services sourced will have been 'local' than would be the case if only the city were considered.

These, however, are not the only values embedded in their analysis. There is no attempt to assess the socio-cultural or environmental impacts of AAG conferences, or any consideration of the negative economic impacts beyond the fact that AAG conferences result in poor-quality jobs being created. These are important omissions. For example, since the opportunity costs of such events and the problems resulting from market concentration and increasing competition between localities are not considered, little in the way of lessons or guidance for AAG conference policy-makers is offered. At present, for instance, the AAG deems it unnecessary to allocate conferences to locations on the basis of their impacts on the local communities, paying little heed to their potential as a regenerative tool. Neither is there any consideration of whether indigenously owned facilities should be used as the venue for the conference rather than TNC-owned facilities (for example Hyatt-Regency, Marriott) so as to increase local multiplier effects. Nor does the AAG encourage locally owned and controlled hotels and businesses to be used by conference delegates, for example by advertising them in the pre-conference brochure, which further reduces the local multiplier effect of the event because of the lower local inter-linkages of externally owned facilities and the fact that profits are exported out of the locality. If such negative impacts had been considered, then detailed policy proposals would have been more forthcoming, such as: that future AAG meetings could carry advertisements for locally owned hotels and eating establishments so as to increase the local multiplier effects; that indigenously owned facilities could be used to host the conference; and that the AAG could be held in relatively deprived communities so as to facilitate their regeneration.

IMPLICATIONS FOR POLICY TOWARDS TOURISM

In sum, this chapter has highlighted how, similar to other consumer services, tourism is witnessing an increasing concentration of ownership and control and is rapidly growing, albeit in such a manner that it is reinforcing, rather than mitigating, many of the spatial inequalities produced by other economic sectors. Moreover, although this consumer service operates as a basic industry, assessing its precise economic impacts has been shown to be a demanding task. Not only are there difficulties in deciding the proportion of total expenditure which is directly attributable to a tourism event and whether expenditure by the local population should be included, but determining the indirect economic impacts has been revealed to be a highly subjective process. How the area is defined, the multiplier chosen and how the multiplier effect is calculated all significantly influence the eventual findings. If assessing the indirect effects is troublesome, then the intangible impacts of tourism are even less amenable to investigation and consequently, are often omitted from economic impact analyses. Many studies, furthermore, and as shown in the case of Randall and Warf's (1995) analysis of the impacts of AAG annual conferences, also fail to consider the full range of negative impacts. This leads to problems since such economic impact analyses then have only limited value as policy-making devices.

Consequently, although the potential of tourism as a tool for local economic development has been realized to a greater extent than in any other consumer service industry, there are a number of ways in which it could more fully contribute to revitalizing local economies. These are here extrapolated from the above analysis. First, there is the matter of spatial policy towards tourism. The state has a pivotal role as a coordinator of the structure and organization of tourism.[7] At present, however, the majority of tourist expenditure remains concentrated in London, and this is increasing over time (see chapter 14). Little drifts out to other localities. So, although there is a putative dispersal policy, this needs to be much stronger if tourism is to be diverted away from London to peripheral areas. This could be achieved by either giving greater emphasis to marketing facilities in other areas or relocating key facilities and events outside London. For example, the Millennium Exhibition, which is intended to attract 30 million visitors, is likely to be awarded to London despite being an ideal opportunity to use tourism to facilitate regeneration in a peripheral location. So too is the new national stadium. Tourism, therefore, is as much in need of a spatial policy to prevent it from reinforcing patterns of uneven development as are manufacturing and producer services.

There is also a need for central, regional and local state agencies to shift away from simply using tourism to generate external income and towards preventing leakage of tourist income out of their area. Unless this is achieved, its impacts on net income will not be maximized. On the one hand, this can occur by encouraging greater local sourcing of the goods and services supplied to the tourist industry, such as by encouraging 'buy local' campaigns to develop a distinct local product rather than create 'identikit' resorts and by facilitating the development of cooperatives of local producers to service the tourist industry (Freitag 1994). On the other hand, it can be fostered through greater local ownership and control of tourist developments and associated industries since these tend to source locally to a greater extent than those which are externally owned and controlled.

It is not solely the optimization of the positive economic impacts of tourism, however, which need to be addressed. Greater consideration is also required of its negative impacts. At present, sociocultural tensions arise because facilities are created which, although desired by tourists, are not those in demand by the local population. Products need to be created, therefore, which meet both local and tourist demand if socio-cultural tensions are to be avoided. In addition, this will have economic benefits. It will reduce the leakage of discretionary local expenditure which might otherwise seep out of locality, an issue which is often overlooked by tourist authorities, and will also improve the quality of tourist employment offered by overcoming the seasonality aspect of this industry (Hohl and Tisdell 1995, Roche 1992).

Finally, and on the issue of the intangible impacts, there is a need for many public-sector agencies to clarify to themselves whether tourism is primarily a lead or supporting sector in their localities. This influences whether they intend to market their locality principally to tourists or to potential investors since place marketing and tourism marketing are distinct entities, often involving quite different target markets and marketing channels and distinct products and place images. Promoting an image for investors and government may be very dissimilar to promoting it to tourists (Bramwell 1993). At present, however, the two are often confused, leading to inadequate strategies for both tourism and inward investment marketing.

In sum, there is little doubt that this consumer service is not a dependent activity. It is a basic industry which generates significant amounts of external income for local economies. The problem, however, is that current local economic policy mostly fails to see tourism as an integral part of an overall economic development strategy which attempts to increase net income for a locality. The result is that endeavours to prevent leakage of tourist income by creating more inter-connected local economies are seldom considered and the full potential of tourism as a tool for economic regeneration rarely realized.

6

SPORT AND THE REJUVENATION OF LOCAL ECONOMIES

INTRODUCTION

Sport is becoming an increasingly popular vehicle for facilitating local economic development, with many localities seeking to attract both 'hallmark' and regular sporting events as well as stadiums. The aim of this chapter, therefore, is to evaluate the benefits to local economies, as well as some of the pitfalls, of pursuing economic rejuvenation through this sector. To do this, the first problem is to define what constitutes sports-related activity, since this is difficult to distinguish from the broader category of recreation and leisure. The working definition of sport adopted in this chapter, following Bale (1991), is that sport is activity dealt with on the sports pages of the national daily newspapers. This focus upon high-profile spectator sport, or sport as display, not play (Stone 1971), is here adopted because this type of sport has the greatest ability to produce income-generating effects for a locality. Commodified sport, or what Ley and Olds (1988) refer to as 'heroic consumption', is of importance to millions of people on a daily and regular basis, taking up more media space than any other industry (Bale 1991).

Despite such widespread popularity, little effort has been spent assessing the economic impacts of sport in Britain. In North America, in contrast, where stadiums are frequently developed with public-sector funds and 'footloose' professional sports teams are then sought to relocate in them (Quirk 1973, Rosentraub and Nunn 1977), economic impact assessments are commonplace, conducted by local municipalities to justify their investments and by teams seeking tax concessions or reduced stadium rentals. Consequently, much of the evidence used here is North American in origin. To examine the role of sport in rejuvenating local economies, first, the nature and extent of sports activity in contemporary Britain will be evaluated, second, the contributions of hallmark events, regular sports events and sports stadiums to local economic development will be examined, and third, a case study of Indianapolis will be employed to display the potential local economic impacts of a sports-focused economic development strategy. Finally, the implications of using sports-related activity as a tool for local economic development will be evaluated. This will reveal that although the sports sector is often criticized as being too small to rejuvenate local economies, it is as large as many of the more conventional manufacturing industries focused upon in local economic policy and contributes to local regeneration not only in terms of its direct and indirect impacts but also through its intangible impacts such as on image enhancement and civic pride. Nevertheless, for sports-related development to become a more effective tool in local economic regeneration, several policy alterations will be required. The chapter thus concludes by highlighting these changes.

THE MAGNITUDE OF THE SPORTS SECTOR

The sports sector in Britain employs 0.95 per cent of all employees in employment (see Table 6.1). As such, it is a small sector relative to other consumer service industries, such as retailing and tourism. Nevertheless, the value-added attributable to sports-related activity amounts to 1.7 per cent of UK GDP, which is greater than for a number of manufacturing sectors, including motor vehicles and parts as well as metal manufacturing, and sports-related employment is of the same order of magnitude as the paper, printing and publishing industries as well as postal services and telecommunications (Henley Centre 1992). In Scotland, for example, sports-related value-added at 2.5 per cent of Scottish GDP is greater than for the mechanical engineering industry and 70 per cent of the level of the electronics industry, both industries upon which much attention has been lavished so far as inward investment is concerned (Pieda 1991).[1] Thus, despite the frequent assertion that it is too small to have a significant impact on economic development (for example Rosentraub et al. 1994), its magnitude is similar to that of many of the manufacturing industries which have conventionally been the focus of local economic policy. Unlike many of these

Table 6.1 Regional distribution of sports employment, 1993

	All sporting activities		Operations of sports arenas and stadiums	
	% of employees	LQ	% of employees	LQ
Greater London	0.786	0.82	0.311	0.79
All South-East	0.947	0.99	0.395	1.00
East Anglia	1.121	1.17	0.420	1.07
South-West	0.964	1.01	0.397	1.01
West Midlands	0.775	0.81	0.321	0.82
East Midlands	0.935	0.98	0.381	0.97
Yorkshire and Humberside	0.922	0.96	0.414	1.05
North-West	0.913	0.96	0.344	0.87
North	0.868	0.91	0.364	0.92
Wales	0.980	1.03	0.511	1.30
Scotland	1.207	1.26	0.466	1.18
Great Britain	0.954	1.00	0.394	1.00

Source: Census of Employment 1993

manufacturing industries, it is also a growth sector. Between 1985 and 1990 in the UK, sports-related jobs increased by 22.4 per cent compared with a 6.3 per cent rise in the economy as a whole, whilst the value-added in the sports sector increased by 48.2 per cent in real terms, rising from 1.4 to 1.7 per cent of GDP (Henley Centre 1992).

However, and as Table 6.1 displays, this consumer service industry is not evenly distributed across regions. Sports employment is concentrated in both the richer core regions of Britain such as East Anglia and the South-West and also in Scotland and Wales. Superficially, conventional demand-led terms could be used to explain its preponderance in the affluent regions, but this does not justify the higher employment in Scotland and Wales. To some extent, this can be argued to be due to the greater preponderance of employment in their 'national' stadiums and arenas. In Wales, for example, 52 per cent of sports employment is in stadiums and arenas compared with just 41.2 per cent in Britain as a whole (see Table 6.1). However, although this factor explains the greater concentration of sports employment in Wales relative to Britain, it still does not fully justify the higher proportion employed in sport in Scotland. Neither does a comparison of the sports-sector studies conducted in the Welsh (Centre for Advanced Studies in the Social

Sciences 1995, Henley Centre 1990), the Scottish (Pieda 1991) and the UK economies (Henley Centre 1992) illuminate why Scotland is so over-represented in sports-related employment. For the moment, therefore, the greater supply of, and demand for, sport in Scotland must remain unexplained, however unsatisfactory that may seem. The important point is that this consumer service industry is not spatially distributed in the manner expected if it were merely a by-product of the wealth generated in other sectors of the economy.

Neither does this sector create mostly low-quality jobs and represent a cost to the public purse, as is commonly assumed. On the one hand, employment is not mostly unskilled and part-time. Indeed, 25.6 per cent of sports jobs are managerial and professional (33.0 per cent for the UK economy as a whole) and just 31 per cent are part-time compared with 28 per cent of all UK jobs (Census of Employment 1993). On the other hand, and contrary to popular prejudice, central government makes a net profit from sport. Income accruing to central government from sports-related activity (£3.56 billion) exceeds spending on sport (£553 million) in 1990 prices by a ratio of 6.7 to 1, compared with a corresponding ratio of 4.5 to 1 in 1985. This is not due to decreased expenditure by central government on sport over this period: spending increased by 13.8 per cent. It is because of the rise in tax revenues generated from sport. Local government, nevertheless, is a net spender. Expenditure by local government in the UK is £1.33 billion, compared with an income of £906 million from sport (Henley Centre 1992).

So, given that this sector is growing, is of a similar magnitude to other industries which are focused upon in local economic policy, and creates professional and full-time jobs to about the same degree as other sectors, the next section evaluates the contributions that it can make to local economic development.

THE CONTRIBUTIONS OF SPORT TO LOCAL ECONOMIC DEVELOPMENT

The local economic impact of sport can be defined as the net economic change in a host community resulting from spending attributed to a sporting event or facility (Turco and Kelsey 1992). Here, the impacts of three types of sports-related activity will be evaluated in turn: regular high-profile spectator sports events;

one-off or hallmark events; and stadium developments. In each case, and as for the tourism impact studies discussed in the last chapter, three types of economic impact will be evaluated: the direct, indirect and intangible impacts. Before commencing this analysis, however, it is important to note that most of the impact assessments discussed below are prospective rather than retrospective, frequently focused upon only one locality, stadium or team and mostly North American in origin (Baade and Dye 1988a). This is not a reflection of any bias on the part of the author. It is simply because this is the information available on the subject.

Regular spectator sporting events

Regular high-profile sporting events by professional teams are both basic activities, directly generating external income for localities in the form of non-local spectators, and leakage preventers, in that they retain expenditure in the area which might otherwise go outside. Schaffer and Davidson (1984), for example, reveal that between 1966 and 1984, the Atlanta Falcons football team contributed $353 million in business activity, $269 million in household incomes and $18.6 million in local government revenue to the Atlanta economy. Page (1990), meanwhile, reports that in Philadelphia in 1983, professional sports had a direct economic impact of $202 million and an indirect impact of $343 million, supporting 757 full-time and 2241 part-time jobs. To display the harmful effects on a local economy of a sports team's withdrawal, moreover, Johnson (1986) estimates that the 1981 baseball players' strike cost Boston $18, 000 in taxes and $650, 000 in spending for each game not played, whilst Miller and Jackson (1988) calculate that $1 million worth of expenditure was lost during the 1980s football strike from in-stadium sales alone.

However, such results need to be treated with the utmost caution. Take, for example, the relocation of the San Francisco Giants in 1992, when it seemed probable that they would leave Candlestick Park in San Francisco, for a new stadium in San Jose. As Crompton (1995) shows, in San Francisco, which was likely to lose the team, it was reported that they contributed just $3.1 million to the city (Corliss 1992), whilst the Mayor of San Jose, trying to persuade the city's residents to approve a referendum allocating $265 million of public funds to a new stadium for the Giants, announced that they would deliver to San Jose 'somewhere between $50 million and $150 million a year in economic benefits' (Fimrie 1992: 52). With such widely varying assessments of the impacts of a professional team, it is little surprise that sports-sector economic impact studies have been treated suspiciously.

This is because studies often choose the multiplier to suit the findings they desire. As discussed earlier in relation to tourism in general, sales multipliers tend to be used by those desiring a high impact, whilst those wishing a low impact (for example when a team has decided to leave) employ income multipliers since household income, as mentioned before, is only one of five potential recipients of the money spent on a sales transaction. Furthermore, studies preferring a high impact often define the area widely so that multiplier effects are larger whilst studies wishing to show a significant impact on the overall local economy define the area narrowly. Similarly, if a high impact is wanted, employment multipliers frequently assume that all existing employees are fully utilized, so a rise in external visitor spending will inevitably lead to an increase in employment levels, whilst this is not assumed by those wishing a low impact (Crompton 1995).

Some economic impact studies, moreover, include local spectator spending to identify high impacts whilst others exclude them (for example Burns and Mules 1986; Crompton 1995; Getz 1991; Smith 1989). In practice, however, only the money which would have been spent outside the locality should be included. This is because some non-local spectators may have been planning a visit to the locality anyway but will have changed their timing to coincide with the event. These 'time switchers' should not be included as generating external income attributed to the event (Crompton 1995, Regan 1991). Van der Lee and Williams (1986), for example, find that 21 per cent of visitors to the Adelaide Grand Prix were time switchers. In addition, some tourists will already be in the locality, lured by other attractions, and will have elected to attend the event rather than do something else. These 'casuals' should also not be included in estimates of external income generation due to the event, except where the event has increased length of stay (Crompton 1995, Regan 1991). Another problem, as Burns and Mules (1986) identify, is that if a public entity contributes $1 million to a $3 million project, then a third of the resulting economic benefits should be credited to them, and not all the benefits. This is seldom done. Frequently, moreover, the negative impacts are ignored (Crompton 1995). For example, there is scant regard paid to the opportunity costs: what could have been produced if the money had been invested elsewhere (Hunter 1988).

Besides these problems involved in measuring the direct and indirect impacts of regular sports events, there are also difficulties in assessing the intangible impacts. It is usually asserted that attracting a well-known team changes the status of a locality by creating a 'major league' image, which captures the attention of those making business location decisions. The purported consequence is that the locality will then grow in areas completely unrelated to sports (Okner 1974). However, and as Baade and Dye (1988b: 42) state, 'This sort of claim is easy to make but hard to validate.' Few studies have been successful in showing whether or not this is the case.

A further related intangible impact is suggested to be that it provides free advertising as a result of coverage of the team. This is again asserted to promote the locality in a positive way. To estimate the value of such media exposure, the City of Chicago Department of Economic Development (1986), for instance, calculate that the championship-winning 1985–6 Chicago Bears football team produced publicity equivalent to a $30–$40 million promotion campaign. This represents 3.4–7.9 per cent of the baseball team's impact on the city economy when measured relative to the direct and indirect impacts. Although some conclude from this that the intangible impacts are unlikely to prove decisive in determining whether or not to adopt a team (Baade and Dye 1988a), they omit to consider other more immeasurable intangibles such as the impacts of championship-winning teams on civic pride.

Civic boosterism may be difficult either to measure or to value, but it cannot be ignored. Sport, after all, aids the development of a sense of identity, loyalty and allegiance to an area (Aldskogius 1995, Maguire 1995). As Churchman (1995: 13) argues, football is a 'major vehicle through which a town or city expresses its aspirations and sense of civic pride' (Churchman 1995: 13). Lever (1973), for instance, reports that when Sao Paulo's most popular team won, production in the city improved by 12 per cent but when it lost accidents increased by 15 per cent, whilst Derrick and McRory (1973) identify a similar tendency in Britain in the context of Sunderland's FA Cup final win, as does York (1991) in relation to Bath rugby club. In a bid to measure such psychic effects, Pollner and Hollstein (1985) assert that if sports success really does make people feel better, then this should result in fewer visits to psychiatric health facilities. No such relationship was identified. This is hardly surprising since there is a major difference between enhanced civic pride and resolving mental health problems.

Assessing the impacts of regular sporting events by professional teams, therefore, is a precarious business. Indeed, ultimately, it is perhaps only the residents who can decide whether the improvement to the locality resulting from the attraction of a team is worthwhile. This is why this issue is increasingly put to a local referendum in the US. When a referendum is conducted, however, most localities usually perceive the positive economic benefits to outweigh the negative impacts. This is also the case for 'hallmark' events, which not only have much greater positive and negative impacts but also fiercer competition to acquire them.

Sporting hallmark events

Hallmark events are international phenomena which function as basic activities by attracting a large number of visitors to the area. They can be either regular but with relatively long time intervals between them (for example the Tour de France) or in different locations each time they are held (for example the World Cup, the Olympics). Often, they are also associated with major infrastructure developments. The revenue collected from the external visitors to such events is substantial. For example, the external income from the 1984 Olympic Games was equal to about 1.6 per cent of the entire 1984 GDP of Los Angeles County (Economics Research Associates 1984), whilst the Seoul Olympics of 1988 generated £2300 million (Bale 1991). The Tour de France, meanwhile, which takes place for three weeks each July, with 950 million television viewers, a £12 million budget and an entourage of 3500 personnel, generated £15 million when in 1994 it visited Britain for two days at a total cost of £1.5 million (National Heritage Select Committee 1995). Indeed, Portsmouth alone witnessed an increase in direct and indirect expenditure of £7.86 million (Portsmouth City Council 1995). Smaller hallmark events can also have major impacts. Blake et al. (1979), examining the 1978 Open Golf championship held at St Andrews, estimate that a total of £3.2 million was spent in the locality.[2]

Nevertheless, although the direct expenditure of tourists tends to be greater than for regular sporting events, the problems associated with measuring the economic impacts are similar. So far as direct external income generation is concerned, expenditure is often counted that would have occurred anyway, such as by 'time switchers' or 'casuals' (Burgan and Mules 1992), and multipliers are chosen and used in ways which conform to the impact funders' desires. For instance, and as for regular sporting events, many studies of hallmark events survey the spenders rather than the recipients (Burns *et al.* 1986, Centre for Applied Business Research 1987, Department of Sport, Recreation and Tourism 1986, Portsmouth City Council 1995). Such sales multipliers, as aforementioned, result in larger impacts than if household income multipliers had been examined. Other studies, moreover, translate expenditure into the number of jobs created using sales/employment ratios in different sectors even though calculating job generation is not so simple, as displayed in the context of the Adelaide Grand Prix (Arnold 1986, Burns *et al.* 1986). Similarly, studies differ in terms of whether they include or exclude all local expenditure in their economic impact assessments. In reality, some local expenditure is important. If, for example, a proportion of the local spectators refrain from going overseas because of the event, then this spending should be counted. It has prevented money from leaking out of the local economy which would otherwise have done so.

The major economic impacts of hallmark events, however, are usually considered to be the intangible impacts. These relate to place marketing and civic boosterism. So far as the former is concerned, by creating a 'major league' image for the locality, hallmark events are viewed as a means of attracting capital and people (Harvey 1989a). The larger the event, the greater the potential for significant positive (or negative) impacts. Take, for example, Johannesburg's bid for the Olympic Games or the 1995 Rugby World Cup in South Africa. Given that the sports boycott was one of the most noticeable aspects of apartheid sanctions, these sporting events can be interpreted as bringing South Africa to the attention of the world as a potential place for investment and as an 'acceptable' tourist destination. Similar intentions lay behind Manchester's Olympic 2000 and Commonwealth Games bids (Leatham 1993). With such benefits to be gained, the competition is fierce to attract such events. Ritchie (1984), however, cautions that because hallmark events are captured essentially to enhance the awareness, appeal and profitability of a destination for investment and tourism in both the short and long term, there is a need to explore their economic impacts long after the actual event. Few, if any, studies do this.[3]

Besides the impacts on a locality's image to outsiders, there are also intangible impacts in the form of civic boosterism. The international attention bestowed on the host locality is asserted to have positive local psychological benefits, manifesting themselves in greater civic pride, self-confidence or a festival atmosphere. Evidence of such impacts, however, is anecdotal. For example, following Sunderland's FA Cup final win over Leeds United in 1973, it was asserted that local output increased, workers were more diligent and industrial relations improved (Derrick and McRory 1973), although no indication of the length of time for which this occurred is recorded. There is indeed little solid evidence so far available which measures this intangible impact.

The impacts of hallmark events, nevertheless, are negative as well as positive. There may be major environmental impacts, as was the case with the Winter Olympics in France (Clapp 1987), and such mega-events also have enormous opportunity costs. Kidd (1979), for example, notes that the attraction of the Commonwealth Games to Edmonton in Alberta resulted in 6000 people being displaced from their homes and that the Montreal Olympics in 1976 resulted in massive municipal debt and the postponement of housing, environmental and public transport projects and reductions in social service budgets. Indeed, Colorado opted not to bid for the 1976 Winter Olympics on the grounds that it would increase developer pressure (Cox 1979). Many impact studies, however, do not take account of such drawbacks. Consequently, although hallmark events have positive and negative economic impacts, difficulties exist in attempting to assess them precisely. It is a similar case when stadium developments are examined.

Stadium developments

Following an earlier vogue in the US (Baade and Dye 1990), the 1980s and 1990s have witnessed an unprecedented level of stadium development in Britain (Churchman 1995, Page 1990). This has included not only indoor arenas (for example the National Indoor Arena in Birmingham, Sheffield Arena and London Arena) because of the growth of indoor sports (Page 1990), but also outdoor stadiums. Churchman (1995), for example, notes that 45 of the 92 football league clubs either erected new stands or moved to new grounds in the early 1990s, in major part instigated by the Taylor Report into ground safety following the 1989 Hillsborough disaster (Home Office 1990).

In the US, stadiums tend to be built and owned by local governments which seek footloose professional sports teams as tenants to play in them (Baade and Dye 1988a, Johnson 1989, Quirk and Fort 1992).[4] In Britain, in contrast, most stadiums remain privately owned by the clubs who play in them (Churchman 1995). Few local authorities fund either stadiums or professional sports teams and next to no teams perceive themselves as footloose. The consequence is a dearth of economic impact analyses of stadium developments. Instead, those which exist are mostly US in origin, undertaken to assess the worthiness of the investment of public money (Baade and Dye 1990, Baim 1990, 1992, Euchner 1993, Rosentraub and Swindell 1991, Zimbalist 1992). As for sporting events, these examine the direct, indirect and intangible impacts of stadium developments.

In an economic impact study of the Louisiana superdome, built at a cost of $122 million, Ragas et al. (1987) estimate the total direct economic impact as $934.28 million between its opening in 1975 and 1984, about two-thirds of which was generated by sports events. This excludes the indirect impacts of the stadium. Real estate impacts which can be attributed to the superdome, such as land value appreciation, meanwhile, totalled $149.4 million, and tax revenues from the stadium over the ten-year period were $115.95 million. Cumulative impacts attributed to the Louisiana superdome between 1975 and 1984 thus totalled about $1.2 billion. The total state subsidy received over this ten-year period was $92.3 million. In non-deflated dollars, the result was that $13 was generated for every $1 invested by Louisiana taxpayers to support its operations and retire its bonded indebtedness. As Ragas et al. (1987: 238) thus conclude, 'Competition among cities for NFL sports franchises and cities' willingness to invest in major capital projects and to provide inducements for the professional teams appears to be economically efficient.'

However, and as with sporting events, there are many problems with assessing the economic impacts of stadiums (see Bale and Moen 1995). Frequently funded by stadium promoters, these studies tend to overestimate the benefits and understate the costs and risks to the taxpayers. The multiplier effects, in particular, are prone to wide variation, and many do not consider the opportunity costs. There are also problems with measuring the intangible impacts. Take, for example, the impact of a stadium on the image of a locality and thus inward investment. Baade and Dye (1988a and b) argue that for eight metropolitan areas, there is virtually no evidence to support the hypothesis that stadium construction led to a difference in manufacturing activity in the period after the stadium had been built. However, given the integrated nature of consumer services (Rosentraub et al. 1994), it might have been more useful to explore its impact on this sector or the service sector more generally. This was not done. Similarly, few studies evaluate the extent to which civic pride is enhanced by such developments.

In sum, although both sporting events and stadiums are high-profile developments which provide direct, indirect and intangible economic impacts for localities, assessment of their precise impacts has been plagued by difficulties of measurement, with value judgements involved at each stage of the analysis. Therefore, to evaluate the potential of the sports sector when it becomes the focus of local economic policy, as well as the difficulties involved in assessing its impacts, the next section examines one such city which has adopted a sports-led regeneration strategy.

THE ECONOMIC IMPACTS OF SPORTS-LED REGENERATION: A CASE STUDY OF INDIANAPOLIS

Indianapolis is one of the few localities to have formulated an economic development strategy for its downtown area which relies substantially on sport (Kotler et al. 1993, Rosentraub et al. 1994, Schaffer et al. 1993, Squires 1989). In 1970, Indianapolis was a city with a declining jobs base, a deteriorating downtown core and an indifferent national image. Consequently, in the early 1970s, an informal group of business, professional, governmental and civic and philanthropic leaders was organized to plan and promote the regeneration of Indianapolis (Schaffer et al. 1993). Its strategy was to rebuild the urban core through sports-related developments so as to prevent a doughnut pattern of prospering outer areas surrounding a deteriorating centre.[5]

In 1977, the strategy was launched. With sports as a central component, a \$2.76 billion construction programme commenced to change the physical character of downtown Indianapolis. This strategy relied not on one team or facility but a series of investments. From 1974 to 1992, over 30 major development projects for the downtown area were initiated, eight of which were directly related to the sports image Indianapolis was trying to establish. These included the Market Square Arena for the Indiana Pacers (basketball), a sports centre, a track and field stadium, a velodrome, a natatorium, the 61000-seat Hoosier Dome for the Indianapolis Colts (basketball), which has hosted the NCAA men's basketball finals three times, a tennis stadium and a National Institute for Sports and Fitness. By 1989, a total of seven national organizations (Athletics Congress of the USA, US Canoe and Kayak team, US Diving Inc., US Gymnastics Federation, US Rowing, US Synchronized Swimming and US Water Polo) and two international organizations (the International Baseball Association and the International Hockey League) had moved their governing offices to Indianapolis. The sports-related investments amounted to $172.6 million or 10 per cent of the total downtown capital investment. If the three hotels, the mall and the station are added, since their success was seen to depend upon attracting people attending sports events, then the total rises to 37 per cent (Rosentraub et al. 1994).

To analyse the impact of Indianapolis's sports-focused downtown redevelopment strategy, Rosentraub et al. (1994) compare: sports-sector job growth in Indianapolis with changes in sports employment in all other metropolitan areas; changing job and salary levels in Indianapolis with nine other cities identified as competing with Indianapolis for development; and growth of the Indianapolis downtown area compared with other cities, since a specific objective of the sports and downtown strategy was to retain and expand employment in this area. Starting with the growth in sports-related jobs, Rosentraub et al. (1994) examine two periods: 1977–89 and 1983–9, as many of the capital projects were not completed until 1982/3. Between 1977 and 1989, numerous cities actually surpassed the expansion in Indianapolis of sports-related jobs. However, if the analysis is limited to 1983–9, when the developments were completed, the increase in sports-related employment compared to all other US cities was impressive, displaying a 60 per cent rise. Only one city did better: Columbus. No city, moreover, surpassed Indianapolis in terms of its growth in sports-related payrolls. Indeed, Indianapolis's growth was almost twice that of the second-best city.

Despite this, in 1989, all sports-related jobs accounted for just 0.32 per cent of all jobs in Indianapolis (0.29 per cent in 1977) and sports-related payrolls for just 0.5 per cent of the total payrolls of all businesses (0.24 per cent in 1977). Hence, the strategy increased the proportion of sports-related jobs by just 0.03 per cent. However, this does not account for the effects on other sectors of the economy. Schaffer *et al.* (1993), using an input/output model, suggest that the investment of $164 million (in 1991 dollars) yielded a return of $683 million in personal income to the residents of Indianapolis and an average annual rate of return of 64 per cent on investment over a 15-year period (Rosentraub *et al.* 1994). In sum, the sports strategy can be accepted as having an impressive rate of return. If the growth of the Indianapolis economy from 1977 to 1989 is compared with growth in nine comparable cities often seen as competing with this city, it can be seen that the number of jobs in Indianapolis grew by 32.9 per cent, exceeded only by Columbus and Minneapolis (Rosentraub *et al.* 1994). Finally, and given the focus upon using sports to develop the downtown area, did the strategy retain jobs in this area of the city to a greater extent than elsewhere? Although by 1990 there was a net decline in the number of jobs downtown compared with 1980, the number was higher than in 1970. Hence, although the strategy could not stop the shift outwards of jobs, it was able to stabilize the employment situation in the downtown area and was more successful in achieving this than many other comparable cities. As Squires (1989) argues, without the sports strategy, Indianapolis would have declined even more.

Consequently, if success is measured in terms of the rate of return on investment, then Indianapolis was successful, but if it is measured in terms of growth in jobs and payrolls, then as Rosentraub *et al.* (1994: 237–8) conclude, 'A sports strategy, simply put, is too small to change the economic patterns in a region.' Nevertheless, not only are they looking at the *regional* impact of one city's downtown area strategy when asserting this, but neither do they explore whether alternative industrial sectors could accomplish more or less. The sports sector, after all, is of a similar order of magnitude to many other sectors more conventionally adopted by local economic policy. The difference, however, is that in an era of increased competition across localities for more traditional sources of investment, the market in sport is still relatively open. As a result, an increasing number of localities are recognizing the benefits of capturing a higher share of this sector. Cleveland, for instance, has pursued a similar strategy of developing new downtown playing facilities for its basketball and baseball teams and Jacksonville, Memphis and Charlotte have all used sports as a cornerstone of their economic development strategies (Norton 1993). In Britain, meanwhile, Sheffield has been the main city to have pursued a sports-focused economic development strategy (see chapter 11).

IMPLICATIONS FOR POLICY TOWARDS SPORTS DEVELOPMENT

At present, the potential of sport as a focus for economic regeneration efforts remains relatively untapped by economic development agencies in Britain. Take, for example, the external income-generating potential of sport. Not only is much less effort put into attracting the sports sector than manufacturing or producer services, but even when events and facilities exist in a locality, attempts to maximize their potential income are few and far between. Take, for example, the national sport of soccer. Visiting supporters are either limited or even banned and are often ushered directly into the ground and prevented from spending in the locality. If football is to integrate more fully into local economic development, then such a culture will need to change in recognition of the shift in the nature of the people attending soccer matches. As Baade and Dye (1988a) note in the US, planned commercial corridors through which fans are channelled in order to maximize local spending as well as ancillary projects (for example sports centres, hotels, convention centres) are increasingly employed.

Besides developing sports-related activity to generate external income, there is also a need to pay more attention to its role in preventing the leakage of income out of localities. In Wales, for example, it is estimated that in 1993, £10.9 million in expenditure was induced into Wales by the sports sector but £170 million was spent outside of Wales on sports-related goods and services, resulting in a net loss of around £160 million (Centre for Advanced Studies in the Social Sciences 1995). Various strategies can be pursued to prevent such seepage. First, local sourcing of the goods and services used by sports events and facilities can be encouraged. Given that 44 per cent of employment in the sports-related economy is in commercial non-sports industries which supply the sport sector with goods and services (Henley Centre 1992), there remain many opportunities for local sourcing and growth in sectors which appear to be superficially unrelated to sport. To achieve this, however, the inter-dependencies between sport and other economic sectors must be recognized. At present, the common supposition is that sport is dependent on other sectors of the economy rather than a means of bolstering these other sectors. Second, greater local ownership of both the events and facilities themselves and their suppliers, including franchises within stadiums and the surrounding area, must be encouraged since this both keeps profits within the locality and enhances the multiplier effects due to such businesses having a tendency to source to a greater extent locally. And third and finally, events and facilities can be provided which the local population would otherwise travel outside of the locality to use.

In this context, there is also a need to consider the social equity issues surrounding economic development based on sports promotion. It is well known that the relatively affluent make better use of sports facilities than the relatively poor. Even the people's game of soccer, which one UK Minister of Sport referred to in the 1980s as a 'slum sport for slum people', is now dominated by relatively affluent spectators. In designing a sports strategy, therefore, it is not just the local residents who need to be given as much priority as outsiders, but also the relatively poor within localities. This has important policy implications. For example, closing local facilities in favour of large 'flagship' facilities, and promoting private sports clubs over public ones, is not conducive to social equity. The trend, however, appears to be towards greater social inequality, with both flagship facilities gaining greater preference in strategies and the commercial sports sector growing faster than the non-commercial sports sector (Pieda 1991, Henley Centre 1990, 1992) exemplified by a 25 per cent fall in local government spending on sport between 1985 and 1990 (Henley Centre 1992).

Finally, a strategic national-level spatial policy towards the sports sector is required. Until now, stadium development, for example, has not been guided by a national strategy (Page 1990), even when the location for a new national stadium was considered, which again is likely to be Wembley in London. There is a case to be made, however, for using both new stadiums and events as a spatial policy tool for reducing regional inequalities and bolstering

89

economic development in peripheral localities. Sport, therefore, although only a small segment of the consumer services sector, is a rapidly growing sector of the economy which appears to have major economic impacts in the localities in which it has been adopted as an economic development vehicle, as displayed in Indianapolis. At present, nevertheless, few British localities have seriously entertained its use as a principal motor of local economic development.

7

UNIVERSITIES AND LOCAL ECONOMIC DEVELOPMENT

INTRODUCTION

By 1993, some 374200 employees (1.8 per cent of the workforce) were directly employed in higher education in Britain, which is more than are employed in agriculture, forestry and fishing combined (Census of Employment 1993).[1] The traditional caricature, nevertheless, is one of 'town and gown' in which university ivory towers are only loosely connected with their localities. The aim of this chapter is to evaluate critically whether this is indeed the case. To do this, and as in previous chapters, the character of this sector and its geographical distribution in Britain will be examined, and then its direct, indirect and intangible local economic impacts will be analysed. This will reveal that universities, which comprise the major component of the higher education sector, are not detached institutions which bestow little on their surrounding areas but, instead, contribute both directly and indirectly to the vitality and viability of their localities. However, for their contributions to local economic development to be strengthened, the final section argues that changes are required in both the external and internal operating environments of universities. The chapter concludes by outlining these necessary alterations.

THE CHARACTER OF THE UNIVERSITY SECTOR

Today's universities have three key missions: generating new knowledge; transferring knowledge to future generations; and serving the needs of industry and the community. However, these missions receive varying emphasis in different types of university, which influences their impacts on local economic development. In this regard, five types of university can be distinguished (Goddard *et al.* 1994). First, there are 'traditional' universities (for example Oxford, Cambridge, St Andrews, Durham) located in areas whose image is very much synonymous with the university, despite these institutions being essentially non-local in orientation and the 'town and gown' tensions being probably greater than in any other type of university. Second, and at the other extreme, there are the ex-poly-technics, created in the late 1960s and early 1970s by amalgamating formerly separate colleges and institutes mostly established to serve local industry and business. Although these ex-polytechnics were taken out of local government control and became 'new' universities in the early 1990s (for example Sunderland, Coventry, Derby), they maintain their strong local links, which are clearly articulated in their mission statements. Third, and lying between the traditional and new universities, are the metropolitan 'red-brick' or 'civic' universities, often established with support from municipal authorities in urban locations, which view themselves as possessing a strong regional vision (for example Newcastle, Sheffield). Fourth, there are the out-of-town campus universities, which, although having fewer local and/or regional roots than the 'red bricks', endeavour to develop them (for example Kent, Sussex, Bath), and finally, there are the place-traversing institutions with a strong regional identity which has been reinforced through the development of decentralized campuses (for example Anglia, De Montfort).

As Table 7.1 reveals, the majority of old universities (i.e. tradi-tional, red-brick and campus universities) define themselves as international institutions seeking to provide local support, whilst no new universities regard themselves in such terms. Instead, new universities see themselves as serving the local community and developing international strength: 74 per cent regarding links with the locality as a high priority compared with just 47 per cent of old universities (Goddard *et al.* 1994). Nevertheless, most universities have some local orientation. Indeed, only four of the 65 universities responding to the survey by Goddard *et al.* (1994) fail to refer to their locality and/or region in their mission state-ments and 81 per cent of all universities attest to increasing involvement with the locality in recent years. The way in which this local orientation is manifested, however, differs significantly between new and old universities. As Goddard *et al.* (1994) identify, new universities focus upon supporting 'regional institutions and industry' (75 per cent compared with 32 per cent of old univer-

Table 7.1 Local orientation of universities: by type of university

	% of universities identifying with strategy		
	Old	*New*	*All*
A community-based institution serving the needs of the local area/region	0.0	11.1	4.6
An institution seeking to contribute to the local area and also develop international strengths	18.4	74.1	41.5
An institution seeking to contribute equally between international research and support to the local area	18.4	14.8	16.9
An international research institution seeking to provide support to the local community where it does not conflict with international research excellence	55.3	0.0	32.3
An international research institution with no particular ties to the local area/region	2.6	0.0	1.5

Source: Goddard *et al.* (1994: Table 7.1)

sities), whilst old universities largely perceive themselves as contributing 'to the cultural life of the region' (71 per cent compared with 44 per cent of new universities). This reflects the greater prevalence in old universities of cultural facilities (for example museums, theatres, art galleries) as well as their contrasting value systems.

The character and extent of their local orientation also differs by geographical location. Universities to the north of the Severn/Trent axis and in large metropolitan areas give greater priority to local and/or regional linkages and supporting regional institutions and industry than southern universities and non-metropolitan universities. Universities in London, meanwhile, put least emphasis on serving the locality. However, this is not the only way in which geography matters.

THE GEOGRAPHY OF THE UNIVERSITY SECTOR

As in many other consumer services, higher education employment is unevenly distributed across space. On the one hand, there are

93

Table 7.2 Higher education employment: % of total employment by region, 1993

Region	Higher education employment (000s)	% of regional employment
All of South-East	133.6	1.9
East Anglia	19.5	2.5
Greater London	62.2	2.0
South-West	25.3	1.5
West Midlands	30.1	1.5
East Midlands	28.7	1.9
Yorkshire and Humberside	30.7	1.7
North-West	42.0	1.8
Northern	15.5	1.4
Wales	15.5	1.6
Scotland	33.3	1.7
Great Britain	374.2	1.8

Source: Census of Employment 1993

significant inter-regional variations, ranging from 1.4 to 2.5 per cent (see Table 7.2). The highest shares of regional employment are in the core regions of the South-East, Greater London and East Anglia, the last being due to the presence of large universities in Cambridge and Norwich. The North, the West Midlands, Wales and Scotland, in contrast, have below average proportions of higher education employment. The result is that this sector reinforces the general regional disparities in the British economy. The South-East, with 33.1 per cent of all jobs in the British economy, possesses 35.7 per cent of all higher education employment. Greater London, likewise, has 16.6 per cent of all higher education employment but just 14.6 per cent of total British employment.

On the other hand, there are intra-regional disparities. Historically, higher education employment was highly concentrated in a relatively limited number of specialized centres. Over time, however, the various waves of university formation have cascaded higher education jobs down the urban hierarchy (see Tight 1996). In the nineteenth century, universities were established mainly in the largest industrial cities of the Midlands and the North of England (for example Birmingham, Leeds, Manchester, Newcastle, Sheffield), whilst the next wave during the 1950s and 1960s endowed mostly smaller towns with such institutions (for example Bradford, Brighton, Lancaster, Norwich, Reading, York, Exeter).

Following this, in the late 1960s and early 1970s, the foundation of polytechnics (which became universities in the early 1990s) meant that many cities with a long tradition of one university now had two (for example Sheffield, Leeds, Newcastle) and numerous new university towns were established (for example Plymouth, Bournemouth, Portsmouth, Derby). The result is that today, few large cities or sub-regions do not have a university.

University employment, nevertheless, remains relatively concentrated in higher-order urban areas. Not only do the 10 largest cities, with 33.9 per cent of all jobs, contain 46.4 per cent of university employment, but 31 of the 33 Local Labour Market Areas (LLMAs) with over 100000 jobs all contain at least one university and an average of 2.4, whilst the 247 LLMAs with fewer than 100000 jobs have only 16 universities between them (Goddard et al. 1994).[2] Indeed, in many university towns and cities, these institutions are often amongst the largest employers. For example, universities are the third and fifth largest employers in Newcastle, third and tenth in Hull, third in York, third and sixth in Sheffield, fourth and sixth in Leeds, seventh in Bradford, eighth in Nottingham and second largest in Huddersfield (Goddard et al. 1994, Leigh 1995).[3] Given this, it would appear that universities possess the ability to make major contributions to the vitality and viability of their local economies.

THE ECONOMIC IMPACTS OF UNIVERSITIES

As in the cases of tourism and sport in the previous two chapters, the economic impact studies which have been conducted to evaluate the contributions of universities to local economic development explicitly conceptualize this consumer service industry as a basic activity which generates external income (for example from central government, local authority fees and research income) rather than as a dependent activity and following this, examine their direct, indirect and intangible economic impacts.

Direct impacts

Although the focus of most of these studies is their local economic impacts, it is salient to mention briefly that universities are also basic activities for the national economy. Examining the direct foreign income generated for Britain by such institutions, the

Committee of Vice-Chancellors and Principals (1995) estimates that the 100000 international students in the UK have an aggregate impact on GDP of between £906 million and £1.2 billion, which generates 35000–53000 jobs. Indeed, these foreign export earnings can form a large proportion of a university's total income. The University of London in 1992/3, for example, received 25.3 per cent of its revenue from overseas students (Goldsmiths' College 1995), which is a higher 'export propensity' than even the business services sector in the City of London, which derives just 21 per cent of its income from abroad (Daniels 1995a). Unlike other sectors, furthermore, universities make a significant contribution to the balance of payments since relatively few UK students go abroad to study. Despite this, studies of the net national income from the university sector are notably lacking.

Instead, and as mentioned, the economic impact studies almost invariably concentrate on their contributions to local economies. During the late 1960s, these were undertaken to explore whether public expenditure on universities might reduce regional imbalances, whilst the post-1980 studies, in contrast, have been conducted mostly to defend universities from cut-backs in public spending. As Table 7.3 shows, the direct impacts are significant. To take just two examples, the nine universities in Yorkshire and Humberside, with a combined income of £728 million, provide 24056 direct jobs (Leigh 1995), whilst the four universities of Greater Manchester generate £468 million and directly employ 12532 (Robson et al. 1995).

Indirect impacts

To calculate the indirect economic impacts of universities, and again as for the impact studies of tourism and sport, most surveys reported in Table 7.3 employ Keynesian multiplier analysis. Here, therefore, the variables which determine whether a high or low impact will be found are not rehearsed again, for they have already been covered in some detail in chapters 5 and 6. Instead, it will suffice to say that these studies use predominantly employment multipliers despite such multipliers being the least reliable of all.[4] They are employed in part because of the relative simplicity of the method and the ease of availability of the data but also because of the fact that they paint universities in a more positive light than, for instance, household income multipliers. Given that most local

Table 7.3 Employment impact of universities: by definition of area

Study	Definition of local economy	Employment Impacts		
		Direct	Indirect (and Induced)	Total
Stirling University	Parts of Stirling and Perth	850–1690	1600–3100	2450–4790
Newcastle Polytechnic	City of Newcastle	1350	2038	3388
Yorkshire and Humberside HEIs	Yorkshire and Humberside	24056	35200	59256
University of East Anglia	Not explicit	800	1000–1200	2857–3057
Bristol Polytechnic	Avon County	1300	1828	3128
Coventry Polytechnic	Local authority	1580	820	2400
Wolverhampton Polytechnic	Wolverhampton and surrounding areas	1331	596 (+169)	2096
Warwick University	Coventry postcode area	2350	550–85 (+290–320)	3190–3520
Southampton University	'Southampton and region'	2438	1619 (+856)	4913
Durham University	Durham City, County and Tyne and Wear	1352	1453	2805
Lancaster University	Lancaster and Morecambe	1352	351	1703
Liverpool Universities	Merseyside	6285	—	—
Strathclyde University	Scotland	2270	3110	5380
Manchester Universities	Greater Manchester	12532	1246 (+1909)	15687
University of Buckingham	Buckinghamshire	239	155	394
Polytechnic of the South Bank	London	1657	1111	2990
Goldsmiths' College	South-east London	850	189 (+222)	1039

Source: Updated from Goddard et al (1994: Table 2)

economic impact studies of universities have been conducted by the institutions themselves in order to display their positive contributions to local economic development, it is hardly surprising to find that the employment multiplier effect has been most widely adopted.

On the whole, the resulting employment multiplier tends to range from 1.1 to 1.8 across these studies (i.e. each direct job generates between a further 1.1 and 1.8 jobs in the locality). Take, for example, the study of the University of Lancaster by Armstrong (1993), which is one of the more detailed attempts to calculate the multiplier effect. In this relatively small locality with a high degree of leakage, it is estimated that an injection of £37.7 million of university expenditure in 1987/8 led to an additional £30.79 million of local income, a ratio of 0.82. Using national accounts data, by sales/employment ratios, this suggests that an initial total of 1352 direct jobs results in an extra 351 jobs, a multiplier of 1.26. If commuters are excluded, the direct employment effect is reduced to 1121 and the extra jobs to 335 or a multiplier of 1.08. A study of Bristol Polytechnic, in contrast, suggests that between 2100 and 2400 full-time equivalent (FTE) jobs were created in addition to the 1414 direct jobs, a ratio of between 1.6 and 1.8 (Taylor 1989). In major part, this higher employment multiplier arises because Bristol has a larger labour market and so less leakage occurs. In both cases, nevertheless, the extra indirect jobs amount to a further 1.0 per cent of total local employment, bringing the proportion of the workforce in these localities employed in these universities to 5.9 per cent and 2.9 per cent respectively (Goddard et al. 1994).

For many, however, this conventional approach for assessing the economic impacts of universities is insufficient. By examining solely their impacts in terms of expenditure, it is argued, a number of more intangible ways in which universities facilitate local economic development are neglected which, as the previous chapters have shown, can be important with regard to the impacts of such consumer services. For this reason, the social accounting or social cost-benefit approach has been increasingly adopted to study university impacts (Bleaney et al. 1992, Goddard et al. 1994, Leigh 1995, OECD 1987, Public Policy Centre and SRI International 1986, Tym 1993, University of Southampton 1991).

Intangible impacts

The more intangible impacts of universities mainly relate to the various supplementary ways in which universities lead or support local economic development. These include the support provided to local companies in the form of technology and non-technology transfer, the growth of new firms emerging from university activities and the additional income generated by the existence of universities in a locality. Here, a brief review of each is provided.

Technology transfer is the process by which a piece of technology in a codifiable form is sold, licensed or otherwise passed on from one party, the university academic, to another, a firm (Dill 1995, Goddard et al. 1994, Matkin 1994).[5] Much dissemination, moreover, is non-technology transfer, whereby existing knowledge is exchanged in the form of improved management awareness. On the one hand, this may take the form of consultancy, such as through management schools, town and regional planning and language departments. On the other hand, it may take the shape of personnel exchanges such as through project work with students (for example on sandwich courses), teaching company schemes or simply through graduate retention in the local labour market. Indeed, for many universities, the widest point of contact with the locality numerically is not through the 'standard' full-time student but through short-course activities, even though they may not amount to much in full-time equivalent terms. In Manchester, Robson et al. (1995) report that there are annually over 100000 individuals taking shortcourses. Indeed, technology and non-technology transfer are often highly interdependent, as the decision to invest in new technology depends upon management awareness. The growth in this form of transfer by universities is displayed by the fact that between 1983 and 1993, old university income from industry rose by 116 per cent in real terms, reaching £131.3 million by 1993/4 (Times Higher Education Supplement, 3 November 1995). This contributes to local economic development by improving the competitive advantage of local companies which engage in such technology and non-technology transfer.

Second, there are the contributions of universities to their local economies through firm 'spin-outs'. This process has a long history and is becoming more common (for example Chrisman et al. 1995), largely as a result of the enormous success of specific past spin-outs, such as Hewlett-Packard from Stanford and Digital

Equipment from MIT (Castro *et al.* 1987). The outcome has been the widespread growth of both campus companies, where the university is a major stakeholder, and science parks (Castells and Hall 1994, Dalton 1995, Florax 1992, Massey *et al.* 1992). Indeed, there are now many studies which attempt to measure the local economic impacts of this spin-out firm formation from universities (Blumenthal 1994, Chrisman *et al.* 1995, Doutriaux 1991, Haug 1995, Mansfield 1991, Segal, Quince and Wicksteed 1985, 1988, Smilor *et al.* 1990).

A third intangible impact of universities lies in their ability to generate additional external income for the locality. On the one hand, this occurs where universities play a supporting role in helping to provide the conditions required for inward investment. Place-marketing strategies, for example, use universities as magnets to attract technologically advanced inward investment by advertising their contributions to the skills base of the local economy and are employed to portray a locality as having a high quality of life so as to attract the 'new service class'. On the other hand, additional income may be derived from tourist spending. Again, universities can play a supporting role through their architecture, museums, art galleries and public spaces, which may add to length of stay and total tourist spending rather than being the prime reason for a visit. However, there are notable exceptions, such as Cambridge and Oxford, where they play a lead role. In Cambridge, for example, tourists spend an estimated £180 million per annum and sustain 5000 jobs (Cambridge City Council 1993).[6]

A further intangible contribution of universities to their localities arises out of the shift in local economic development towards partnerships. In many localities, universities have taken on a greater role in the new 'quangocracy' of coalitions which dominate local decision-making. Indeed, given that much funding for new projects is often conditional on partnership formation, universities have increasingly played an important and often lead role in levering such external income. This, however, is not their only contribution to local governance. Individual university staff are also involved in a range of partnerships, including Urban Development Corporations, Training and Enterprise Councils, school governorships and various trusts. Nevertheless, few universities keep a record of such activity by staff.

Besides these intangible economic impacts, universities also contribute in a number of ways to social and community life, such as by providing local sporting facilities, libraries, cultural life, continuing education and community services. Many universities, for example, provide multi-art centres, and student unions are often important to local youth culture, providing a niche between club and major venues. In addition, 63 per cent of universities have theatres and 59 per cent concert halls open to the paying public, 78 per cent hold free public lectures (17 per cent charge), lists of experts are provided to the media, and 90 per cent of universities offer some public access to libraries as well as the provision of exhibition space, language centres, bus services and sports facilities (Goddard *et al.* 1994). They also make important contributions in the realm of volunteering. For example, 125 student action groups provide 15000 volunteers who work with existing voluntary and statutory community organizations. The estimated economic value of this activity is £6.9 million (Goddard *et al.* 1994).

Consequently, there are a wide range of intangible contributions which universities make to their localities besides their more direct and indirect economic impacts. As in the case of the study of tourism and sport, however, these are easy to define but difficult to measure. Nevertheless, at least some attention has been paid to these intangible impacts in previous studies. The negative impacts of universities and the ways in which universities are shaped by (rather than just shape) localities receive far less attention.

Negative impacts and inter-dependencies

The vast majority of economic impact studies of universities, most of which are conducted by the institutions themselves, tend to focus their attention on the benefits of their universities. Far fewer examine the negative impacts. Indeed, although some analyses have started to pinpoint these issues, especially in terms of the social and environmental impacts such as in the realms of rented accommodation and increased noise and traffic pollution (for example Armstrong *et al.* 1994), discussion of their negative impacts is much less developed than in studies of tourism and sport. This reflects both the infancy of university impact studies as a field of inquiry and, as in the sports sector, funders' aspirations. However, it is not only the negative impacts which are under-played.

Most of the literature on the contributions of universities to local economic development, as shown above, tends to emphasize their external income-generating abilities and how they lead revitalisation. However, the connection between universities and their localities is not the one-way relationship painted in these impact studies. Universities not only shape, but are also shaped by, the vitality and viability of the other sectors of the economy. Technology and non-technology transfer, to take just one example, are not a one-way passing of knowledge from the university to business as portrayed in many studies but, in reality, a two-way flow of knowledge. For many universities, moreover, their success in attracting students both determines, and is determined by, their nightclub sector in an iterative process. They are also shaped by the overall image of their localities. Consequently, although the dependent-activity perception of universities has to a large extent been overcome, it has been replaced by a 'lead'-sector caricature, which has caused some loss of understanding of the complex inter-dependencies which exist between universities and their localities. For this reason, these inter-dependencies will be returned to in chapter 11.

POLICY IMPLICATIONS FOR THE UNIVERSITY SECTOR

This analysis, nevertheless, has revealed that universities appear to make significant contributions to their local economies. They function as basic-sector activities which generate external income and jobs and often also lead and support local economic regeneration by enhancing the image of the area, inducing inward investment, facilitating spin-out companies, improving the skills base, encouraging technology and non-technology transfer and contributing to the quality of life through their social and cultural facilities. Here, however, it is argued that for these contributions to local economic development to be strengthened further, changes are required in both the external and internal operating environments of universities.

Commencing with the external operating environment, the first alteration needed is in how universities are conceptualized by existing and potential funders. Like many other consumer services (for example sport – see chapter 6), central government views universities as a cost to the public purse rather than an economic investment (Chrisman *et al.* 1995). As Jones and Vedlitz (1988) reveal in the US, however, there is a strong correlation between the quality of a state's university system measured by grants received and state aggregate income, income per capita and business creation. They conclude that higher education plays a significant role in a state's economic fortunes. Leaving aside the issue of whether the quality of the university sector is a cause, an effect or inextricably inter-related to a state's wealth, the important point is that this rethinking of the role of the university sector in economic development is in its infancy in Britain. Therefore, the first wave of research on the local economic impacts of universities now needs to be succeeded by a second wave of studies on the national economic impacts of both individual universities and the university sector as a whole. Without such a reconceptualization, funding seems likely to continue to decrease and with it, university impacts on local (and national) economic development. There is also a need to change the way in which universities are perceived by both other public-sector funders and the private sector. Although universities are moving ever further away from being a free state service, there appears to be a process of lagged adaptation so far as many potential funders are concerned, who remain entrenched in the view that universities can provide 'free' research and consultancy services.[7]

It is not only the conceptualization of universities by actual and potential funders which needs to alter. A rethinking of the spatial planning of the university system is also required. Given that the university sector, much like tourism and sports, discussed in the previous two chapters, currently operates as a form of aspatial public policy which has significant unintended spatial impacts, there is perhaps a case to be made for stronger central planning at a regional and sub-regional level of the university system. At present, however, government cannot and does not use the university sector as a regenerative tool, such as by investing more in universities in peripheral localities or by deciding to relocate or expand universities in these areas. Although some limited bottom-up initiatives are occurring by universities themselves, such as by linking up with further education colleges in peripheral localities, a top-down view is missing in relation to the spatial planning of university provision.

If universities are to contribute to a greater extent to their localities, a change will also be required in how quality in universities is assessed, especially in terms of research. At the time of writing, one of the few ways for individual universities to gain a greater slice of total government funding is by achieving high grades in the Research Assessment Exercise (RAE). The current rules of this competition, however, severely curtail academic research which is principally of benefit to the locality of a university. The 1996 RAE ranked departments, in major part, on the 'best' four publications of each of their staff. To gain an 'international' rating and thus relatively more funding, much time was spent by academics publishing academic books and articles in (little read) leading international academic journals. Other forms of research output, such as consultancy reports solely for a local authority or business, were less valued. So too were articles in professional or trade journals (which are often widely read). The result is that academics have been encouraged to write and conduct research for publication in prestigious international academic journals and to reduce time spent either resolving local problems or disseminating knowledge locally, causing a shift away from the applied and professional arena.

There is little doubt, moreover, that new universities, which have more locally oriented and applied mission statements, have been disadvantaged in the RAE relative to the old universities since the criteria are based more on the missions of the latter (see McKenna 1996). As Mohan (1994) notes, the problem for new universities is that if they attempt to improve their research rankings, they may lose their strong local grounding and links with community organizations, whilst old universities face the problem that pursuing research outputs which conform to these criteria prevents them from developing strong local links, which may harm their identity. Because they are driven by intense competition for such research funds, the resulting problem is that universities are discouraged from fostering local links.

For universities to contribute more actively to their local communities, therefore, a change in RAE policy is badly required. To enable the pursuit of new knowledge generation of international importance whilst encouraging greater contributions to be made to local development, one solution might be for future RAEs to rank natural and social science research in the following manner. Highest rated could be research which both generates new knowledge of international importance and applies (or at a minimum, discusses the implications of applying) this new knowledge to the solving of 'local' problems (i.e. research which 'thinks globally and acts locally'). Next most highly rated might be research that solely 'acts locally', transferring existing knowledge to solving local problems, and lowest rated could be 'blue skies' research, which generates new knowledge but fails to consider how it could be applied locally. Whether or not this is the most appropriate framework for encouraging research which links globally oriented new knowledge generation with 'local' concerns, it is obvious that the current quality assessment criteria and measurement methods are far from ideal for encouraging academics to devote anything other than minimal time to engaging with their localities in a positive developmental manner.

To enable universities to interact more fully with their localities, changes are required not only in their external but also in their internal operating environment. A first practical step is for universities to adopt a local sourcing policy for their goods and services wherever possible. This will automatically lead to higher local multiplier effects. Of course, and as in any other sector, consideration will need to be given to issues of cost, quality and reliability of supply (Williams 1994a), but the principle of buying locally, all other things being equal, is one which will bring major benefits to the local economy.

More widely, and as Goddard *et al.* (1994) argue, if universities are to increase their impacts on local economic development, then mechanisms for better structuring university/community relations will need to be developed. They argue that, first, an audit of existing community linkages is required and, following this, each area of activity (for example economic development, built environment, and social and community development) needs to be given a clearly identifiable socket within the university into which individuals and organizations can plug. Such a university/community development strategy, they argue, should be informed by analysis of the university's capacity as well as of community needs. The problem at present is that few universities have begun to structure relationships with their communities in such a systematic manner. Greater coordination, therefore, is required if university interactions with the locality are not to remain a kaleidoscope of idiosyncrasy and complexity. Currently, for example, only just over half of all universities have regular meetings with local authorities, with just 43 per cent of old universities recording good relations (81 per cent of new universities). Similar contrasts apply to university relations with other local bodies. Only 38 per cent of old universities meet regularly with other agencies such as TECs/LECs, compared with 50 per cent of new universities. Neither are there formal mechanisms in many universities for monitoring and promoting such relations. Indeed, only 58 per cent of new universities and 32 per cent of old universities actively encourage this involvement (Goddard *et al.* 1994).

A possible problem for many reading the above paragraphs might be that if universities respond to local needs and conditions, this will conflict with their increasingly global mission of seeking new knowledge and translating and passing this information on to others. There are few instances, however, where thinking globally cannot result in acting locally so far as the natural and social sciences are concerned, or even the humanities. Although knowledge generation cannot be restricted exclusively to that which is useful to local interests, it could well be argued that it should be applied, tested, researched, generated or transferred locally rather than externally wherever possible to maximize the impacts of universities on their localities.

If British universities adopt this development path, they will not be alone, and neither will they be in danger of losing their globally oriented mission. In the United States, for example, 50 urban universities have created the Coalition of Urban and Metropolitan Universities. As Lewis (1995) states, this coalition has set out in a common declaration how knowledge creation, dissemination and application can be made more responsive to the needs of their metropolitan areas. Its policy agenda is one which British universities could usefully explore if they are to become more responsive to their communities. These include redirecting teaching towards under-served groups, encouraging practical research on urban issues and strengthening professional service through partnerships with public and private enterprises. The thread linking these actions is a public commitment to make their resources more readily available to local communities. Arising in the context of increasing urban problems and reductions in expenditure on universities, much high-quality cross-disciplinary new knowledge has been generated and encouraged by a re-evaluation of the reward structures for academic staff.

To encourage such a local orientation amongst academics in Britain, a similar process is required. The promotion criteria for staff, which presently closely mirror what is rewarded by the RAE, need reconsideration. As Brian Robson argues, 'there is still a dilemma about blue sky research versus hands-on consultancy. There is still a feeling that money derived from research councils has a higher status than money derived from elsewhere' (cited in Targett 1995: 2). This, however, is not a perception. It is a reality. At present, personal reward and promotion are more likely to follow from publishing in academic journals and winning research council contracts than from producing 'coal-face' consultancy reports. Applying the above-stated 'think globally and act locally' criterion to reward structures, it can be argued that the desired cultural shift of combining globally oriented new knowledge generation with local action would ensue. Nevertheless, such an internal change is largely reliant upon an alteration in the criteria employed in the RAE for it to be successfully implemented.

CONCLUSIONS

In sum, universities are no longer the detached institutions of the 'town and gown' caricature. Instead, they make an active and positive contribution to local economic development both as external income generators and through their contributions to image enhancement, inward investment, spin-out new firm formation, improving the skills base, and technology and non-technology transfer as well as the quality of life through their social and cultural provision. The problem, however, is that aspects of the external and internal policy environment within which universities operate mitigate against closer links being forged with their localities. The challenge facing universities as we approach the end of the millennium is to transcend the notion that they must either 'think globally' or 'act locally' and to develop new ways in which they can do both at the same time. Unless this is achieved, they will be unable to retain their three key missions of generating new knowledge, transferring knowledge and serving their local community, and will under-perform as catalysts for local economic development.

8

RETAILING AND ECONOMIC DEVELOPMENT

INTRODUCTION

More than any other consumer service evaluated in this book, retailing is widely held to be a dependent activity with little, if any, role in economic development. As Lowe and Crewe (1991: 345) put it, 'the retail trade is typically seen as an insignificant backwater, a sector which is somewhat tangential to the "real" world of production'. This is clearly illustrated in local authorities' attitudes towards new retail developments. Although separate agencies have been established in many localities and regions, as well as nationally, to encourage inward investment from manufacturing and producer services, a retailer's application for planning permission is normally met by questions about the trade diversion impacts of the proposed development. Ironically, and unlike all other economic sectors, the more successful the retail development is forecast to be, the less likely it is to be granted planning permission. A new shopping centre that diverts more than 10 per cent of trade away from an existing centre, for example, is likely to be rejected, while one that diverts only 5 per cent is more likely to be accepted. This is not the case with manufacturing and producer services. Quite the opposite. The larger the trade diversion impact, and thus jobs created, the greater is the probability that it will be granted planning permission and even be given enticements to locate in the area (Williams 1996c).

It is not just amongst local authorities that retailing is marginalized from economic development. It is a similar case in academic discourse. In geography, retailing is cordoned off into both retail geography and increasingly cultural geography, with few economic geographers considering it a subject worthy of investigation.[1] In town and country planning, similarly, the retail sector is only considered in retail planning, which is wholly separate from local and regional economic development.

In a bid to counter this isolation of retailing from economic development, first, this chapter examines the nature and extent of the retail sector in contemporary Britain, and second, its contributions to economic development using case studies of regional shopping centres (RSCs) and cross-channel shopping for beer. This will reveal not only that trends in retailing mirror those in many other consumer services but also that this sector contributes to economic revitalization as a basic sector-activity and leakage preventer as well as through its supporting or lead role in enhancing the image of a locality. Finally, therefore, the policy implications of recasting the role of the retail sector in economic development are explored.

THE RETAIL SECTOR IN CONTEMPORARY BRITAIN

The expansion of the retail sector has been a principal feature of contemporary economic restructuring. Between 1981 and 1993, the number employed in retail distribution increased from 2.060 million to 2.136 million whilst the share of the workforce in this sector expanded from 9.7 per cent to 10.1 per cent. By 1993, therefore, retailing employed 6.5 times as many employees as agriculture, forestry and fishing, 2.5 times as many as the construction industry and the same number as the producer services sector, and was two-thirds the size of the manufacturing sector (Census of Employment 1993).

It is a sector, moreover, which is undergoing radical restructuring.[2] One of the most significant changes is in retail ownership patterns. Until 1964, manufacturers could decide the price of their goods in the shops. Efficient stores, eager to sell more, could not compete by charging their customers less. Supermarkets could not undercut village grocers. The abandonment of price fixing in the 1964 Resale Price Maintenance Act, however, meant that small amounts could no longer be bought from a manufacturer for the same price as bulk purchases. By bulk buying, supermarket chains could compete by cutting prices. A retail revolution resulted. The outcome was a shift from local to corporate ownership and an increasing concentration of retail capital.

Between 1960 and 1989, the multiples increased their share of retail sales from 33 per cent to 80 per cent (Warpole 1992). Today, therefore, retailers are some of the largest companies in the UK. Examining the top 100 companies by their market capitalization, *The Sunday Times* (19 November 1995) ranked Marks & Spencer the tenth largest, Sainsbury's twenty-sixth, Tesco thirtieth, Argyll Group fifty-first and Asda Group sixty-sixth. Take, for example, British grocery retailing. Wrigley (1994) shows that between 1982 and 1990, the market share of the top five food retailers – Sainsbury's, Tesco, Asda, Safeway and Gateway – increased from under 25 per cent to 61 per cent of national sales.[3] As Guy (1994) reveals, the outcome was that in 1989, although 90.6 per cent of all businesses were independent single outlets, they commanded just 17.3 per cent of total turnover. Meanwhile, 0.03 per cent of businesses had 100+ outlets and controlled 62.9 per cent of turnover.

Recently, however, competition has come from foreign-owned companies. On the one hand, foreign discount stores, such as Netto and Aldi, have entered the British market with its higher net profit margins than other EU nations (Guy 1994). On the other hand, US-owned 'warehouse clubs' (for example Costco) have attempted to enter the British market, although the alleged heavy lobbying of government by the oligopoly of British retail capital has managed to curtail this invasion (Marsden and Wrigley 1996). The overarching trend in British retailing, nevertheless, is away from local indigenous ownership and towards national and international companies with multiple outlets, which results in an outflow of funds from local economies in the form of profits and income.

Similarly, there are some significant trends so far as its inter- and intra-regional distribution is concerned. On an inter-regional level, the common assumption is that as a dependent activity serving local demand, this sector will be over-represented in core regions since its size reflects the level of affluence. As Table 8.1 illustrates, however, although the affluent core regions of the South-East and South-West are indeed over-represented by retail employment, so too are the peripheral regions of the North and North-West. Yet again, therefore, the popular wisdom that consumer services such as retailing are simply a by-product of affluence is called into question. Much more research is thus required of the factors determining the variable regional levels of retail employment,

RETAILING

including the influence of retail ownership patterns, the geo-demo-
graphics of different regions, the configurations of retail spending
and inter-regional retail expenditure flows before regional variations can be understood more fully.

On an intra-regional level, meanwhile, the conventional notion is
that retail employment will be over-represented in higher-order
urban agglomerations. This is based on Christaller's aforementioned central place theory (Christaller 1966). However, and as
Table 8.2 illustrates, such employment is under-represented in
the highest-order urban areas whilst many lower-order areas, which
are meant to have lower concentrations of retailing, both are over-
represented in retail jobs and have witnessed significant increases
in such employment. To a major extent, this is perhaps a reflection
of population decentralization both from inner-city to suburban
areas and from large cities to smaller surrounding towns. However,
whether population follows retailing or vice versa remains very
much open to question. Neither, moreover, are these trends solely
due to population dispersal. The significant increases in retail jobs
in outer rural areas cannot be explained simply in terms of population increases.

From this, it is evident that the conventional assumptions about
the inter- and intra-regional variations in retailing can no longer be
accepted at face value. Instead, a new finer-grained geography of
retailing must be developed which reflects contemporary reality. To
achieve this, however, and as with so much work on consumer

Table 8.1 Regional distribution of retail employment, 1993

	% of employees	Location quotient
Greater London	9.69	0.96
All South-East	10.35	1.02
East Anglia	9.82	0.97
South-West	10.52	1.14
West Midlands	9.43	0.93
East Midlands	9.81	0.97
Yorkshire and Humberside	9.93	0.98
North-West	10.28	1.02
North	10.35	1.02
Wales	9.80	0.97
Scotland	10.00	0.99
Great Britain	10.12	1.00

Source: Census of Employment 1993

111

Table 8.2 Employees in retailing, by type of area, 1981–91

	1981	1991	Change 1981–91 (%)	Location quotient 1991
Inner London	152.8	142.8	−6.6	0.73
Principal cities	193.4	183.0	−5.4	0.88
Outer London	165.2	177.7	+7.6	1.16
Other metropolitan boroughs	287.3	312.3	+8.7	1.00
Other cities	292.4	308.5	+5.5	1.01
Industrial districts	227.0	275.2	+21.2	0.97
Ports and resorts	131.3	147.9	+12.6	1.18
Districts and new towns	97.7	133.7	+36.8	1.08
Mixed urban/rural	329.3	413.9	+25.7	1.07
Outer rural	175.9	220.3	+25.3	1.00
Total	2049.3	2311.3	+12.8	1.00

Source: Townsend *et al.* (1996: Table 11.5)

services, this activity can no longer simply be assumed to be a dependent activity serving local demand.

THE CONTRIBUTIONS OF RETAILING TO ECONOMIC DEVELOPMENT

The retail sector, far from being a dependent activity, is, like many other consumer services, increasingly contributing to economic development as an external income generator and leakage preventer as well as more intangibly by providing a supportive or lead role in rejuvenation. The following two case studies illustrate this. First, the impacts of regional shopping centres (RSCs) are examined and second, and to display that retailing also contributes to national economic development, the impacts of cross-national shopping for beer are evaluated.

The impacts of regional shopping centres

RSCs are edge-of-town purpose-built developments of at least 500000 square feet of retail floorspace with a wide range of commercial establishments, usually including catering and leisure facilities, which are designed, planned, managed and marketed as a unit (Williams 1994e). In the UK, applications for the construction of RSCs reached their peak in the latter half of the 1980s, with 54 applications between 1982 and 1991 (Reynolds and Howard 1993). At present, nevertheless, just four RSCs are in operation: the MetroCentre in Gateshead (opened in 1986), Merry Hill in the West Midlands (1989), Lakeside in Thurrock (1990) and Meadowhall in Sheffield (1990). Several more are awaiting or under construction at the time of writing: the Trafford Centre in Manchester, Cribbs Causeway in Bristol, Blue Water Park in Dartford and the White Rose Centre in Leeds. It appears, however, that besides these, no more RSCs will be permitted.

This is not through lack of demand. It is because ever more stringent conditions have been set for allowing the building of new RSCs (Department of Environment 1993, 1995). This is because such complexes are asserted to suffer from three problems. In social terms, they are perceived to be poorly accessible, especially for deprived social groups; in environmental terms, they are argued to have negative impacts, especially with regard to 'greenfield' sites and CO_2 emissions; and in economic terms, they are viewed as parasitic activities, which have a negative effect on overall job numbers and destroy city centres, by diverting trade away from existing retail facilities.

Recognizing the potential of RSCs as vehicles for economic development, however, a number of studies have challenged both the evidence and logic used to support these assertions (Williams 1992a and b, 1993, 1994e, 1995). First, RSCs have been shown not to suffer from poor access. Indeed, they are rated by shoppers as highly on the issue of 'ease of access' as other shopping centres, and they have invested so heavily in public transport infrastructure that customer profiles reveal that lack of access to a car makes no difference to whether RSCs are visited (Howard and Davies 1992). The result is that the clientele is relatively representative (Howard and Davies 1992). Indeed, some perceive social integration, rather than social exclusiveness, as the principal function of RSCs (Glennie and Thrift 1992). Second, not only have RSCs been shown to be built on 'brownfield' rather than 'greenfield' sites, but the only data on the resource consumption and pollution impacts of RSCs are revealed to be those constructed for public inquiry purposes for a consortium of local authorities opposed to the building of the Trafford Centre in Manchester (Williams 1995).

Third, both the empirical data and the logic underlying the

notions that RSCs are parasitic activities which have a negative effect on overall job numbers and destroy existing city centres by diverting trade away from current retail facilities are highly questionable. The only evidence mustered in a recent literature review of their negative employment impacts is derived from forecasts of RSC proposals (Whitehead 1993). None are 'real' statistics of what has happened in practice. Similarly, the data used to reveal that RSCs transfer jobs rather than create them are controversial. One study uses employment change in large and small grocery retailers, which are asserted to be predominant in RSCs and city centres respectively (Whitehead 1993). Nevertheless, few RSCs contain large grocery stores and the notion that small grocery retailers are predominant in city centres is based more on a longing for some past golden age than the reality of the high street today. Hence, the data are inappropriate for the task.

Not only are the data inadequate, but so too is the logic. According to Planning Policy Guidance note 6 (PPG6), the key issue for a new retail development is whether it will deleteriously affect the 'vitality and viability' of existing facilities (Department of Environment 1993, 1994, 1995). Retail impact studies thus analyse its trade diversion impacts (for example Howard and Reynolds 1989). This is the principal policy focus because it is deemed that retailing is merely a 'dependent' activity, circulating money within a locality. It is clear, however, that RSCs are not. They create jobs in the same way as any other basic-sector activity, namely by generating external income so as to stimulate growth in the local economy. There is therefore little reason for using different economic impact assessment methods from those applied in other sectors. Here, therefore, and in a bid to assess the economic impacts of the MetroCentre in Gateshead, rather than using retail impact assessment techniques which provide little scope for evaluating the more positive outcomes of such developments, the standard economic impact assessment techniques used in previous chapters which measure the direct, indirect and intangible impacts are applied to this RSC.

The economic impacts of the MetroCentre

The MetroCentre in Gateshead opened in 1986 on a site previously used for dumping ash from the local power station. To provide a preliminary and tentative measure of its contribution to the local economy, the degree to which it induces income into the locality and thus jobs, and prevents the seepage of income out of the locality can be examined. Retail activities are frequently perceived, as outlined above, to be dependent activities merely circulating money within the local economy. However, the MetroCentre, with a turnover estimated at £200–40 million per annum in 1988 prices (Howard and Reynolds 1989), attracts 78 per cent of its trade from people whose home is over 5 km from the MetroCentre (Howard and Davies 1987) and thus outside Gateshead. The amount of external income induced into Gateshead by this RSC is therefore £156–87 million. Concomitantly, since there are 5500 jobs in the MetroCentre (Gateshead MBC 1992), approximately 4300 are extra jobs which would not otherwise have existed in Gateshead. Few manufacturing inward investment schemes, it should be noted, have generated this number of jobs in any locality in the UK.

These, moreover, are merely the jobs within the RSC itself. Excluded are the many additional jobs created, including: the direct jobs in associated activities outside the RSC which can be attributed to it (for example from visits to nearby tourist attractions); the indirect jobs which result in the chain of suppliers of goods and services; and the induced jobs, resulting from successive rounds of expenditure out of incomes produced from direct and indirect employment. Here, no attempt is made to calculate the gross number of jobs, or the net employment impact of the MetroCentre for the region as a whole. This is for two reasons. First, even if a net job loss were to result, this is no reason for the MetroCentre to be opposed. This would be due to productivity rises in the retail sector which, if they occurred in other sectors, would be applauded, not decried (Williams 1996c). Second, it is improper to criticize the job gains in Gateshead because of the losses elsewhere, as is all too common (for example Whitehead 1993). It is inconsistent to embrace, on the one hand, endeavours to promote manufacturing growth poles, yet on the other hand, oppose a retail development which performs the same function. To reject retail job gains in Gateshead because of losses elsewhere is not a relevant criticism when such 'inter-locality specialization' is extensively and energetically fostered in all other sectors of the economy (Cooke 1990: 173). Parity of treatment is required across all sectors of the economy. The grounds for treating RSCs differently from other basic industries are extremely flimsy indeed.

Besides producing additional income for Gateshead, the Metro-Centre also lessens the leakage of income. This is essential, as stated, because income generation alone is insufficient for economic growth to occur. Gateshead MBC (1990) estimate that in 1986, before the opening of the MetroCentre, there was a net loss of expenditure of 24 per cent for durable goods and 21 per cent for convenience goods. In 1988, after its commencement, although Gateshead still exported £80 million (42 per cent) of its durable goods expenditure, the spending of outsiders on durable goods in the Borough through the MetroCentre and Retail World was over £180 million. This endows Gateshead with a net income from durable goods expenditure of £100 million. For convenience goods, meanwhile, the net outflow of expenditure in 1988 was reduced to 3 per cent of total convenience goods expenditure. Therefore, from a position of a net outflow before the inauguration of the MetroCentre, after its opening, Gateshead witnessed a net inflow in terms of durable goods expenditure and radically reduced the extent of the outflow in terms of convenience goods. The MetroCentre, consequently, has provided a powerful stimulus to the economic revitalization of Gateshead by changing retail expenditure patterns in a way which significantly improves the net income of Gateshead.

Furthermore, it cannot be speculated that although retail expenditure is more likely to be kept within Gateshead because of the MetroCentre, subsequent off-flows will increase because of the greater prevalence of multiple retailers in the MetroCentre than in the conventional high street. Of the 300 or so outlets in this complex, some 145 are let to local firms (Chesterton 1993). It is very unlikely that many high streets could equal such a distribution of tenants skewed towards local enterprises. Consequently, the MetroCentre emerges as perhaps less likely to foster the flow of profits out of the area than traditional high streets.

It is not just in its role of generating external income and curbing its seepage that this complex has helped to revitalize Gateshead. Several additional intangible benefits related to its role in enhancing the image of the area as a means of attracting investment deserve to be mentioned and have elsewhere been argued to be common to most British RSCs (Williams 1992a and b, 1993, 1995). Indeed, this was an explicit and critical factor in its being given planning permission in the first place (Barford 1987). Put another way, the MetroCentre has operated as a 'growth pole' by attracting enterprise to cluster around it. On the one hand, this is due to its 'magnetic effect'. Additional businesses have established here to feed off both the RSC itself, in the form of goods and services provision, and its clientele. On the one hand, this clustering is also due to its 'demonstration effect'. It has shown other businesses that firms can succeed in this area and given them greater confidence to relocate than would have been the case if the Metro-Centre did not exist. An RSC which is built at a cost of £350 million and employs 5500 people, putting an investment of approximately £64000 per employee into an area, represents a considerable fixed investment achieved, incidentally, with little or no public-sector leverage. This has bolstered confidence in the locality, resulting in an upward spiral of investment, as witnessed by a clustering of activity. The alternative would doubtless have been continued site dereliction involving a local eyesore, untreated contaminated land and a direct and visible sign of the area's lack of success and its community's impoverishment (Williams 1997). Moreover, since the MetroCentre symbolizes the consumerist ethic which pervades late twentieth-century Britain, this 'cathedral of consumption' has been used to make the area appear a more attractive place for consumers to live in and/or visit.

In sum, the MetroCentre has bolstered the regeneration of Gateshead as a generator of external income and a preventer of income leakage as well as through its growth pole and place-marketing functions. The story would be remarkably similar if Meadowhall (see chapter 11), Merry Hill or Lakeside were examined. Neither is it just edge-of-centre RSCs which have such economic impacts. The same applies to in-town purpose-built shopping centres. Walker's (1995) analysis of Princes Quay in the centre of Hull clearly displays that there are similar economic impacts. Nor is it simply at the local and regional level that the retail sector can play a positive role in economic development. Given its widespread parasitic status, it is here useful to display that retailing acts not only as a motor for local economic development but also as an engine for national growth.

'Booze-cruising': the economic implications of cross-channel shopping for beer

To a limited extent, cross-channel shopping took place for many years prior to the Single European Market (SEM). Holiday-makers would return with foreign foodstuffs and other goods whilst middle-class southern households would occasionally visit France for the specific purpose of buying produce unavailable in the UK. Apparently, no empirical research was conducted into this form of cross-national shopping. It is certain, however, that its nature and extent have changed considerably with the advent of the SEM (Williams 1997, Williams and Windebank 1996a). To examine the changing magnitude and character of cross-channel shopping and its impact on the economies of France and the UK, this section focuses upon the specific issue of cross-channel shopping for beer. Of course, cross-channel shopping for other goods occurs, but the case of beer is one of the few where its extent and impact have been documented, albeit mostly by the UK brewing industry, which obviously has some interest in the issue.

With the arrival of the SEM on 1 January 1993, duty-paid and duty-free allowance restrictions on beer were effectively eradicated. People could bring into the UK 110 litres of beer, paying French rates of duty, which are much lower. Initially, for a 5 per cent alcohol-by-volume beer, French excise duty was about 1 penny per pint, although this was increased to bring it into line with EU minimum rates to 4.2 pence in May 1993. The equivalent beer in the UK has 30.7 pence duty per pint, around seven times more. The limit of 110 litres per person thus means a saving of £51.41 on a 5 per cent alcohol-by-volume beer trip in duty alone. This 110-litre limit, however, is only indicative. If a person can provide an assurance that it is for personal consumption, larger amounts can be imported.

The outcome is that people are increasingly importing beer from France either for their own consumption or for (illegal) sale to third parties. In 1994, imports of duty-paid beer totalled 404 million pints, compared with 333 million pints in 1993, when a further 26 million pints were imported duty-free. This 1993 figure, moreover, is over twice the level of 1992 (Whitbread 1994). By 1994, therefore, duty-paid imports and duty-free purchases equalled 4 per cent (3.5 per cent in 1993) of the UK beer market, or 16.5 per cent (15 per cent in 1993) of all beer consumed in the home (Brewers and Licensed Retailers Association 1994, 1995).

The total loss to UK retailers in 1994 was estimated to be £540 million (£430 million in 1993). Of this imported beer, it is calculated that 60 per cent would have been drunk in a pub or club. This represents a loss in retail sales of some £300 million in 1993, equal

118

to the bar sales of 1500 average public houses. The remaining 40 per cent would have been sold in the off-licence trade, with the loss of another £130 million. These lost beer sales resulted in 1994 in a VAT loss of £80 million (£64 million in 1993) and a duty loss of £121 million (£83 million in 1993), £201 million in total (£147 million in 1993). Meanwhile, the French Chancellor of the Exchequer gained an additional £33 million in excise duty and VAT. The Henley Centre have indicated that with the 'knock-on' effects, more than 10000 UK full-time equivalent jobs have been lost, three-quarters of which will have been in pubs and clubs (Brewers and Licensed Retailers Association 1994). Consequently, because of the UK government's failure to react to the advent of the SEM by reducing the differences in taxes on beer, there has been a significant leakage of income out of the national economy and loss of jobs.

Explaining the cross-channel shopping phenomenon

Whitbread (1994) assert that 63.4 per cent of cross-channel beer purchases in 1993 were made by individuals (10 per cent of whom bought more than 110 litres, some for resale), 29.4 per cent were bulk purchases by van and 7.2 per cent duty-free purchases on ferries. To give weight to their campaign for a reduction in taxes on UK beer, the brewing industry have emphasized the illegal aspect of this trade. However, they provide little evidence that a large proportion of the imported beer is illegally being sold to third parties. As Smith (1994) asserts, not only is it less likely that people will now smuggle more goods than before, but the flexibility of the rules and regulations makes it difficult even to be a smuggler, given the indicative limit and the fact that this can be exceeded if the beer is for personal consumption. It is, as he says, simply easier to feel like a smuggler post-1993.

Indeed, the growth of cross-channel shopping cannot be explained simply as an attempt to engage in illegal activity. Instead, explanations have to be sought in the changing nature of shopping. Given that shopping is now one of the country's foremost 'leisure' pursuits (Lowe and Wrigley 1996, Newby 1993), it is unsurprising that an activity which combines the 'leisure' pursuit of foreign travel with the incentive of saving money on one's shopping should be so popular. The media have tellingly referred to cross-channel shopping trips as 'booze cruises'.

Moreover, cross-channel shopping needs to be seen in the context of the trend towards self-provisioning in shopping. As Gershuny (1978) highlights in his concept of the 'self-service economy', households have increasingly provided services for themselves using capital goods at their disposal, rather than buying formal-sector services on the market. Shopping is exemplary of this process. Consumers increasingly conduct for themselves work previously undertaken by the retailer. This self-servicing activity is not only internal but also external to the retail store. Groceries, for example, are no longer transported by the supplier to the shop located on the corner of the consumer's road. Instead, consumers are obliged to transport their own groceries from large-scale superstores using their own means of transport and their own money and time. Furthermore, storage is increasingly the responsibility of the consumer rather than the distributor (for example home freezing). Indeed, 'real' price reductions in various products over the past thirty years are precisely due to producers and retailers using the unpaid labour of the consumer. Hence, in the 'self-service economy' of retailing, the shopper has become used to buying items in large quantities and storing them for use at a later date, and s/he is accustomed to travelling long distances to pay lower prices. Cross-channel shopping is simply an extension of these trends. Of course, the amounts bought are larger and the distance travelled longer, but the travel costs are off-set by the savings made. It is not, however, simply an economic decision. Consumers engage in such activity for both economic and social reasons (Williams and Windebank 1996a).

Policy implications of the cross-channel shopping phenomenon

To understand the policy implications, it is first necessary to examine who gains and who loses from this growth in cross-channel shopping due to the SEM. The first beneficiary is the UK consumer. The relaxation of import quotas following the SEM means that they can now legally import beer at lower prices than would be paid in the UK. The second beneficiary is the French government and the French economy, especially in the frontier areas. France is enjoying the benefits of increased revenue from the rise in income on duty which is paid on beer purchased in France by UK consumers and the jobs created by such increased expenditure.

The losers, meanwhile, are the UK government and its economy. The government has lost approximately £64 million in VAT and £83 million in duty, and cross-channel shopping for beer adds some £430m to the UK balance of payments deficit (Brewers and Licensed Retailers Association 1994). In particular, this loss of sales has significant impacts upon the UK brewing industry, pubs and clubs and off-sales retailers, especially but not exclusively in the South-East. The Brewers and Licensed Retailers Association (1994) find that although the destination of a significant proportion of total personal duty-paid imports is indeed London (37 per cent) and the South-East (19 per cent), large proportions also go to the North of England and Scotland (19.8 per cent), the Midlands and Wales (19.7 per cent) and the South-West (5.8 per cent). It should be noted, however, that the major UK retailers are not losing. Many, such as Safeway, Sainsbury's and Tesco, have set up outlets in the French channel ports in order to sell beer to UK residents. Tesco, for example, estimate that it lost £35 million in UK retail sales to cross-channel shoppers in 1993 and forecast an £80 million shortfall for 1994. In Calais, there are now 30 supermarket or cash-and-carry operations geared to the cross-channel drinks trade, including Sainsbury's, Tesco and Victoria Wine, and this is also the case in Dunkirk, Boulogne and Cherbourg as well as near the entrance to the Channel Tunnel. The consequence is that UK retailers are not losing market share to the extent one might assume, despite the shift of consumer spending into France.

In France, therefore, beer retailing is performing a basic-sector role. The French government, by keeping its taxes on beer lower than the UK, is generating external income for the French economy and creating new employment opportunities, especially in the relatively deprived northern coastal areas. The UK, on the other hand, is leaking income profusely. Because of its refusal to adjust duties on beer, money which was before spent in the UK is now going abroad, leading to an increase in the balance of payments deficit and meaning that exports have to increase to compensate for this loss.

The outcome was that beginning in 1994, cross-channel shopping became the focus of a concerted campaign by the brewing and tobacco industries, whose objective was to force the UK government to lower excise duties. In the November 1994 Budget, however, the Chancellor of the Exchequer initially refused to reduce taxes on beer. Instead, he froze them, arguing that he was instead going to lobby heavily at EU level for the minimum excise duties on such goods to be raised so as to prevent the French from continuing to undercut the UK. However, following this budget, there was a furore over the government's decision to introduce value-added tax on fuel. The government eventually withdrew this scheme and instead, in part, raised the necessary revenue which had been lost by increasing excise duty on beer by 1 penny. Whether the decision is taken to reduce UK beer taxes, or whether the UK government lobbies heavily for the minimum duties to be raised in the EU, this case study clearly reveals that the retail sector is not a 'parasitic' activity but, rather, has an income-generating function in national economic development. It also displays how it can negatively influence the economy if income leakage occurs. In other words, it provides an example of how retailing can positively or negatively contribute to national economic development.

It is an instance, therefore, of harmonization with the rest of the EU being required to bring the trade balance back into equilibrium. If this is not done, leakage of income will continue so far as the UK is concerned. Cross-channel shopping, therefore, is engendering a 'market-led' harmonization of duty and VAT levels in the EU, in contrast to a concerted and managed harmonization, which was opposed by the UK. As the Brewers and Licensed Retailers Association (1994: 23) asserts,

> The level of duty paid imports is not the result of a Single Market but the inevitable consequence of maintaining unsustainable differences in taxes when frontier controls have been removed. The UK is attempting to determine beer duty according to an isolationist concept of fiscal sovereignty which simply ignores the Single Market.

By overlooking people's willingness to engage in cross-channel shopping, the UK government has overestimated its own ability to stand alone within the SEM. Indeed, it is ironic that a government which preaches the virtues of market forces is being brought to book by those self-same forces, as its citizens take up the call to entrepreneurship and make for the supermarkets of Calais.

IMPLICATIONS FOR POLICY TOWARDS THE RETAIL SECTOR

Given this recasting of the role of the retail sector in economic development, what are the policy implications? Starting at the national level, the case study of cross-channel shopping clearly reveals that central government tax policy is a major determinant of whether a country's retail sector is competitive. Countries are not islands. Indeed, by revealing people's willingness to engage in cross-channel shopping, this case study uncovers that the UK government overestimated its ability to stand alone within the SEM.

Similarly, at the intra-national level, localities perhaps also overestimate the extent to which they are islands. As shown above, retailing can function as a basic-sector activity. Despite this, retailing has received little attention in economic development circles. Take, for example, the currently popular strategy of town centre revitalization. Many local authorities have recently been active in establishing town centre managers (Farrar and Smith 1995). However, few explicitly connect town centre revitalization with their economic development strategy. Indeed, some strategies even exclude it when their authority is endeavouring to raise its town centre up the retail hierarchy so as to attract more external income (Bedwell 1993). Yet to focus upon retail-led revitalization would be a strategy in which there is less competition than in other spheres of economic development. Whilst nearly all local authorities are chasing manufacturing industry and producer services, very few are seeking to develop the retail sector as a prime focus of their economic development strategy. For example, few Unitary Development Plans explicitly propose the construction of an RSC, which, as has been shown, would have major economic benefits for the locality in which it was built.

If such larger-scale income-inducing retail developments are ignored by economic development practitioners, then the more income-retaining activities do not even get noticed. Discussions of either the ownership of retail facilities or local sourcing are nearly non-existent.[4] Few economic development practitioners, for example, would deem it worthy of their attention or scarce resources to aid indigenous retail establishments in their area to compete with multiple chains. Similarly, few would consider it worthwhile to devote resources to encouraging retail enterprises to engage in the local sourcing of their goods. Such income-retaining initiatives, however, are powerful potential shields in the economic development armoury.

At present, even if people buy local goods and services, much of the income still flows out of the locality, for example in the form of profits, on account of the external ownership of much retail enterprise. This, in major part, is a result of the increasing concentration of retail capital, discussed earlier. Yet in contrast to the widespread and vocal debate about the position of retailers on the road out-of-town (Williams 1994c), there has been little reaction to such retail concentration. Indeed, attempts to encourage indigenous retail enterprises have been few and far between.

It is not only a change in the ownership of retail outlets which could stimulate economic development. Initiatives to encourage greater local sourcing of merchandise could also promote growth. However, there is little evidence of such a policy being pursued. Even in a rural economy, where some lip-service is paid to 'buy local' campaigns, few see the retention of a retail outlet which, if it closed, would cause people to purchase the goods externally, as important to its economic development strategy. It is seen more as a social service than an economic initiative. Nevertheless, if retail outlets are locally owned and sell locally produced goods, then there is every reason, because of their income-retaining abilities, that their local economic impacts will be far greater than those which are externally owned and source their goods from external suppliers (see chapter 13).

A pioneering development in this regard is the 'Out-of-this-World' initiative of the Creative Consumer Co-operative Ltd. These stores adopt a policy of sourcing locally a proportion of their merchandise (Creative Consumer Co-operative Ltd 1995). On the one hand, they seek to purchase locally from community businesses. The intention, in so doing, is not only to provide local jobs but also to lessen environmental impacts through reduced transportation. On the other hand, they extend local purchasing by joining Local Exchange and Trading Systems (LETS). LETS are local associations whose members make offers of, and requests for, goods and services and then exchange them priced in a local unit of currency (Williams 1996d, e, f). Members of the local association thus buy goods in the shop using local currency, which the shop then uses to pay both staff and local suppliers. In so doing, it overcomes a problem of conventional local purchasing schemes:

that even if people do buy locally, the business may often be owned externally and/or the goods and services sourced from external suppliers. Using local currency means that this money can only be used to purchase further local goods and services within the system. None of it can leave the locality, nor can it be used to purchase goods and services outside the area, since it has no value outside the local association. The effect is that a 'closed' system is created whereby none of the local money leaks out of the area (Williams 1996e). Consequently, in terms of its policy of local sourcing and using local currency systems to do so, Out-of-this-World provides a positive example which could be replicated by other retailers.

CONCLUSIONS

In sum, this chapter has revealed that the role of the retail sector in economic development must be recast. It is a means of generating external income for an area, preventing the leakage of income out of an economy and supporting or leading revitalization. The MetroCentre, for example, not only generates external income for the area and prevents leakage of income, but also acts as a growth pole and valuable place-marketing tool. The function of the MetroCentre in revitalizing a local economy, nevertheless, is no different from that of any other retail-sector activity. The difference is merely one of magnitude. The case of cross-national shopping, moreover, has shown the role of retailing in broader national economic development and displayed the order of magnitude of the loss which has resulted for beer sales from the advent of the SEM and the failure of the government to react.

As such, economic policy can no longer regard retailing as a dependent activity. It is a vehicle for economic development. However, to enhance its contribution, this chapter has argued that there is a need to promote basic-sector retail enterprises, indigenous ownership and local sourcing of merchandise. Moreover, more academic research on the contributions of the retail sector to economic development is badly required. Relative to the wealth of evidence on other consumer services, there is very little literature on the economic contributions of retailing to local economic revitalization. Given that 10.1 per cent of all employees are employed in this sector, this seems a project worthy of attention.

9

CULTURAL INDUSTRIES AND LOCAL REVITALIZATION

INTRODUCTION

In the United States, cultural industries became firmly embedded in local economic regeneration strategies in the 1970s (Frost-Kumpf and Dreeszen 1995, Whitt 1987), whilst in the UK, this occurred in the 1980s. The result is that cultural industries, in contrast to many other consumer services, are no longer portrayed as an 'extravagant cost' to localities, but rather as an important weapon in their economic regeneration armoury. The aim of this chapter, therefore, is to evaluate the potential contributions of cultural industries to local economic development. To commence, the magnitude of the cultural industries sector will be outlined, along with an analysis of the ways in which they contribute to local economic revitalization. Unlike the previous chapters, however, and because of the considerable number of policy initiatives taken towards the cultural industries, this will be followed by a charting of how cultural industries policy has changed over time, a typology of current approaches and a mapping of how cultural industries policy varies spatially. Finally, the ability of current cultural industries policy to regenerate local economies will be critically evaluated and several modifications suggested in how this sector is used to pursue local economic development.

The term 'cultural industries' was first coined by the Frankfurt school in the 1930s, but now refers to those industries which transmit meaning and which involve creative input (Garnham 1990). This includes not only 'high art' but also forms of creative activity associated with popular culture (Wynne 1992a). As Bianchini (1993a) asserts, the cultural industries cover not only the 'pre-electronic' performing and visual arts (theatre, music, painting and

126

Table 9.1 Employees in selected cultural industries in Britain, 1991

Industry	No. of employees	% of all employees
Film production, distribution and exhibition	23900	0.11
Radio and television services	74500	0.35
Authors, music composers and other own account workers	10900	0.05
Libraries, museums and art galleries, etc.	68600	0.32
Gramophone records and pre-recorded tapes	4500	0.02
Printing and publishing of newspapers	76700	0.36
Printing and publishing of periodicals	20700	0.10
Printing and publishing of books	19500	0.09
Other printing and publishing	214500	1.00
Musical instruments	2100	0.01
Total	515900	2.41

Source: Census of Employment 1991

sculpture) but also contemporary industries like film, video, broadcasting, photography, electronic music, publishing, design and fashion. At the outer edges, it also merges with the tourism, heritage and leisure industries.[1]

To estimate the size of this sector in Britain, Table 9.1 explores the level of employment in those cultural industries identifiable in the 1980 SIC index. Although these sectors do not include all the cultural industries, they still employ 2.4 per cent of all employees, which is more than those employed in the energy and water supply industries and nearly twice as many as are employed in agriculture, forestry and fishing.[2]

To examine the geographical distribution of this industry, Table 9.2 analyses the location quotients (LQs) for four types of cultural industry. This reveals that they are not as evenly spread as might be expected. Instead, they are concentrated in London and its hinterland. Such concentration, moreover, is increasing over time. The English Tourist Board (1991), for instance, identifies that whilst 69 per cent of all heritage-related investment took place in London and the South-East in 1986, by 1991, it has risen to 75 per cent. One outcome is that over half of all visits to cultural tourist attractions in England in 1992 occurred in this core region (English Tourist Board 1993).

It is not only the concentration of cultural industry jobs in London and the South-East which results in spatial disparities, but also the type of jobs and business which are located there.

Table 9.2 Location quotients of selected cultural services in Britain, 1991

	Film production, distribution and exhibition	Employees, radio and television services	Authors, music composers and other own account workers	Libraries, museums and art galleries, etc
Greater London	3.1	3.5	2.9	1.4
South East	1.9	1.9	1.7	1.1
East Anglia	0.5	0.6	0.3	0.7
South West	0.8	0.6	0.6	1.2
West Midlands	0.4	0.7	0.8	0.9
East Midlands	0.4	0.3	0.4	0.6
Yorkshire and Humberside	0.4	0.5	0.5	1.0
North West	0.6	0.6	0.9	0.8
North	0.6	0.4	0.4	1.0
Wales	0.6	0.9	0.8	1.1
Scotland	0.5	0.6	0.8	1.1

Source: Census of Employment 1991

As chapter 14 will show, London houses the headquarters of many major firms in cultural industry production and most of the principal outlets for cultural-oriented consumption. Consequently, it has greater proportions of externally oriented headquarters functions and related higher-level cultural industries than other localities and regions, which tend to have relatively lower-order cultural industries. To a large extent, this is because competition for cultural industries takes place at a strictly hierarchical level. On the one hand, there is fierce international competition between global cities, especially for the higher-order functions in this sector, as is currently witnessed in the battle between Paris, Milan and London to become international centres of culture and fashion. On the other hand, there are a multiplicity of 'second cities' such as Glasgow, Manchester, Liverpool, Bristol, Sheffield and Leeds in Britain, which compete with each other for lower-order functions. Witness, for example, the contest between Leeds and Sheffield for the surplus artifacts from London's Royal Armouries Museum (see chapter 12). Indeed, it is at this level that the vast majority of cultural industry research has been concentrated in both Europe and North America (see, for example, Bassett 1993, Bianchini 1993a and b, Griffiths 1995, Whitt 1987). To understand why such intense inter-locality competition is taking place for cultural industries, it is necessary to understand their contributions to local economic revitalization.

THE CONTRIBUTIONS OF CULTURAL INDUSTRIES TO ECONOMIC REVITALIZATION

As Ashworth and Ennen (1995) argue, art is no longer seen as a product of the wealth generated elsewhere in the economy. Similar to other consumer services, it is viewed as a potential motor for local economic development. The cultural industries are seen as a means of generating external income for a locality, improving its net income, enhancing the image of a place on the mental maps of external investors, cultural tourists and the indigenous population, and having a 'pull' effect on potential investment. Each is now considered in turn.

Cultural industries are viewed as generating external income for localities in two ways: as basic industries exporting their product (for example film, television, painting); and as basic consumer services which attract cultural tourists into a locality in order to spend their money (for example performances, festivals, screenings and concerts in concert halls, studio theatres, temporary circuses and art galleries). In Bristol, for example, it is suggested that 2600 jobs have been directly created in the cultural industries (Bassett 1993). Direct employment, however, is only one part of the equation. As Myerscough (1988) estimates, there are multipliers of 0.33 to 0.5 in the cultural industries sector, meaning that a further 850–1300 indirect jobs will have been created in Bristol. The most extensive locality study of the impacts of cultural industries has been undertaken in Manchester, where it is estimated that in 1988/9, the turnover of the cultural industries sector was £343 million producing 13100 jobs: 9100 directly and 4000 indirectly (Hughes and Gratton 1992).[3]

Moreover, it is not only at a local level that this consumer service industry generates external income. Wynne (1992a), for instance, estimates that in 1986, the publishing industry contributed £550 million and recorded music £495 million to the net national balance of trade in the UK. Myerscough (1988) argues that tourists with an arts ingredient in their trips spent some £3.1 billion in 1985/6 (25 per cent of total tourist earnings), of which £950 million was spent by domestic tourists and £2.2 billion by overseas visitors. These figures cover all trips which involve the use of arts facilities. Not all such spending, however, can be said to be induced by the arts since the trips of some arts attenders are not undertaken wholly or specifically with cultural objectives in mind. On the basis of answers to questions aimed at judging the strength of influence of the arts on the decision to make the trip, he estimates that £2 billion of tourist spending was attributable to the specific influence of the arts, which is 16 per cent of total tourist spending and 27 per cent of overseas tourist spending. These figures relate only to museum and gallery and theatre and concert attendance. A wider interpretation of cultural facilities would doubtless have resulted in larger estimates of the international external income-generating impacts of the cultural industries.

However, the contributions of the cultural industries do not lie solely in their ability to generate external income. As argued in previous chapters, given that net income is the critical factor in local economic regeneration, preventing the leakage of income out of a locality is equally important. In this regard, Myerscough (1988) has shown that local resident spending on cultural industries cannot be viewed as contributing little or nothing to the local economy. Instead, he finds that if such facilities had not been available in the locality, 90 per cent of this money would have been spent outside the area. In so doing, he reveals that cultural provision not only generates external income but, just as importantly, retains income within the locality which would otherwise leak out. Cultural industries, therefore, have an important income-retaining impact. They can ensure that money which appears to be 'ring-fenced' by individuals for this activity is spent locally.

The significance of the cultural industries to economic development, however, extends beyond their impacts on net income. They also have intangible impacts in the sense that they positively reconstruct the image of places on the mental maps of both external investors and the indigenous population. Culture, after all, is the means by which localities communicate their character and distinctiveness, and make positive statements about who they are, what they do and where they are going (Montgomery 1995b). It is, in other words, a major weapon to be wielded by areas wishing to express the 'quality of life' to be experienced by living in them. Indeed, as perceived quality of life has become a more important factor in locational decisions and cultural elements have been recognsized as a key ingredient in delivering this image (Cwi 1982, Rogerson et al. 1988), the cultural industries have taken on heightened importance (McKellar 1988). Griffiths and Williams (1992), in a content analysis of the promotional materials used by local and regional bodies in the UK, find that a quarter gave primary stress to cultural factors in their promotional work, and virtually all agencies gave them secondary emphasis. Hence, the mobilization of cultural resources in place marketing is one of the recent ways in which cultural policies have become a legitimate part of regeneration strategies.

Finally, cultural industries can not only improve an area's net income and recreate the image of places but can also provide intangible impacts through their more direct 'pull' effect. Cultural flagship projects, for example, can act as a magnet to draw further development into an area by acting in a similar manner to a 'key industry' in a manufacturing 'growth pole', as well as providing a demonstration effect that business can successfully relocate in that area. During the past two decades, for example, Paris has witnessed a considerable number of individual spectacular projects, including the Orsay Museum (which opened in 1986), the Grande Louvre project (1989), the Institute of the Arab World, Opéra-Bastille, La Villette, the new French Library and the Grande Arche at La Defense. The impact is that it has retained its position on the world stage through a process of image enhancement, and many of these projects have also had both a magnetic and a demonstration effect (Kearns 1993). Take, for example, the Grande Arche. This is the centre-piece which has enlivened and consolidated La Defense as a location for many global and European headquarters of major multinational companies.

Given this overview of the contributions that cultural industries can make to local economic revitalization and the widespread adoption of cultural industries in economic development strategies, the policies which have been adopted towards the cultural industries will now be examined, commencing with a brief historical excursus concerning how the cultural industries have been treated in different periods.

THE EVOLUTION OF LOCAL CULTURAL POLICY

Many, although not all, of the present-day cultural initiatives build upon a cultural infrastructure of libraries, museums, theatres and concert halls forged in a previous round of local cultural policy in the nineteenth century. These earlier cultural strategies, however, were primarily undertaken for social, not economic, reasons. As Minihan (1977) argues, from the 1830s onwards, support for culture occurred in the context of rising social unrest and was undertaken not only to improve morality and social order, but also to provide a means by which social elites could define themselves and construct new hierarchies of taste (Bassett 1993, diMaggio 1982). Griffiths (1993), in addition, argues that they served to identify a place's position in the political and cultural life of the nation. The cultural industries, nevertheless, were not seen as engines of economic growth. Culture and the arts brought respectability to an area's wealth; they were not the creator of that wealth. Their legacy, however, was a cultural infrastructure that was to be developed and transformed in a quite different context by the cultural strategies of the post-war period (Bassett 1993).

In the decades immediately following the Second World War, a consensus existed that there should be state support for the arts through the mechanism of the Arts Council. In local policy-making, moreover, cultural provision was a non-controversial area. Because there was a narrow definition of culture as the pre-electronic arts, few links were made between an area's cultural resources and their possible use for economic development purposes (Bassett 1993). Instead, cultural policy was essentially a type of welfare service in which the principal concern was to provide access to an artistic and cultural heritage whose social and ideological content was largely unchallenged (Mulgan and Warpole 1992, Bassett 1993, Griffiths 1995).

In the early 1970s, however, cultural policies adopted a more political slant. The new social movements emerging in the post-1968 era (feminism, youth rebellion, environmentalism, community action, gay and ethnic minority activism) saw cultural activity as inextricably related to political action. The result was the development of an 'alternative' cultural sector encompassing experimental theatre groups, rock bands, independent film-makers and cinemas, free radio stations, small publishing houses, radical newspapers and magazines. These challenged the orthodox distinction between high and low culture (Bianchini 1993a). This was taken on board by the New Left strategies, which rejected the cultural elitism inherent in the distinction between high art and popular culture; questioned the assumption of a national culture by supporting the emergence of a plurality of counter-cultures and ethnic minority arts; broadened the concept of culture to embrace new technologies and new cultural forms such as television, film and radio; formulated cultural industry strategies to develop popular control of these new technologies; and politicized culture by linking cultural policies to the new municipal socialism and political opposition to Thatcherism (Bianchini 1993a).

During the 1980s, nevertheless, the strategic objectives of cultural policy shifted from the social and political concerns of the 1970s to economic development priorities (Bianchini 1993a). This occurred for two reasons. On the one hand, the globalization of corporate-profitability strategies resulted in deindustrialization in the advanced economies as multinational firms shifted unskilled and semi-skilled parts of the manufacturing production process to newly industrializing countries (see chapter 2). The result in the advanced economies was increased inter-local competition for inward investment and the consumer revenues of the skilled technical and managerial labour which accompanied it (Dicken 1992, Harvey 1989a and b). On the other hand, there was a shift to the right in the political climate in many advanced economies, which was accompanied by a contraction in local government finance.

If support for cultural provision was to continue, it had to be justified in terms of its economic contributions. Initially, as Bianchini (1993a) explains, this was a defensive tactic to preserve existing levels of cultural expenditure in that the language of subsidy was gradually replaced by the language of investment. As time passed, however, the potential contributions of the cultural industries to local economic development became increasingly apparent. They were recognized as generators of external income in their own right and it became evident that they were also a key weapon for helping to attract inward investment by offering the quality of life package favoured by the new service class (Bassett 1993, Bianchini 1993b, Booth and Boyle 1993, Eckstein *et al.* 1996, Griffiths 1993, Montgomery 1995b, Richards 1996). As Urry (1990) and Bourdieu (1985) argue, unlike the traditional bourgeoisie, this service class does not possess substantial economic capital and instead, derives prestige from its 'cultural capital' by defining standards of fashion, taste and style. The creation of a lively cosmopolitan cultural life was thus increasingly seen as crucial for attracting mobile international capital and specialized personnel, especially in the high-technology industries and advanced service sectors (Bianchini 1993a).

Contemporary cultural industry strategies

More than in any other consumer service industry, it has thus been increasingly recognized that cultural industries directly contribute to economic revitalization both as export industries and as mechanisms for inducing people and investment into a locality. Consequently, most cultural industry *strategies* adopted by localities can be placed on a continuum which has a pure producer-oriented approach at one pole and a pure consumer-oriented approach at the other. Running parallel to this, in addition, is a spectrum of *goals* with an 'integrationist' approach which seeks social cohesion and supports all forms of cultural provision at one end and a 'promotional' approach at the other which focuses upon attracting external people and investment and often relies on flagship projects and established art forms. Here, therefore, and to highlight the contrasting nature of cultural industry strategies, the characteristics of the approaches at the polar opposites of these continua will be outlined.

Those adopting the consumer-oriented strategy develop those cultural industries which attract cultural tourists into the locality in order to benefit from their spending and to entice investment by projecting a better quality of life for professional and executive employees. To a lesser extent, they also attempt to prevent local residents from spending their money outside the locality. Originating in the USA (Whitt 1987), this approach can be seen across much of Europe, including Birmingham (Lim 1993) and Glasgow (see chapter 10). Numerous methods are employed to facilitate such an approach. For example, high-profile events or festivals may be launched so as to encourage cultural tourism and raise the reputation of the locality. Rolfe (1992) identifies over 500 professional arts festivals in the UK which last more than one day and notes their proliferation since the 1980s. Box-office income in 1991 amounted to £17.6 million. Notable examples include the Edinburgh Festival and the Notting Hill Carnival, each attracting 1.5 million people. Other examples include La Merce in Barcelona and First Night in Boston (Schuster 1995).[4] Another strategy is to focus upon 'soft' infrastructure provision in the form of cultural animation. This involves the programming of events and spectacles to encourage people to visit, use and linger in public places and is becoming an increasingly popular means of stimulating urban vitality in many localities in advanced economies (Montgomery 1995a).

A further common method used by this consumer-oriented approach is to reinvent the night-time and evening economy of cities (Bianchini 1995, Montgomery 1995a, Lovatt and O'Connor 1995). Although participation in night-life is not a mass experience in the UK as it is in other European nations (Bianchini 1995), there have been an increasing number of attempts to promote it (for example in Manchester and Leeds). With the recognition of the growing inter-locality competition for segments of the cultural consumption market, such as 'clubbing', and of the fact that the evening economy represents between 5 and 15 per cent of local GDP (Montgomery 1995a), opportunities have been seized to expand market share (see chapters 11 and 12). Problems encountered, however, have been a perception by authorities, such as liquor-licensing magistrates, that the evening economy is something to be regulated and suppressed, rather than developed, and overcoming the population's perception that city centres are only for youth in the evenings. One method employed to overcome the latter has been to delay the closing time of retail outlets so as to build a bridge into the evening economy and retain a cross-section of people in city centres.

In contrast to this consumption-oriented approach, other strategies towards cultural industry development place more emphasis on cultural production. These producer-oriented approaches focus upon developing the commercial cultural industries such as audio-visual firms, publishing and fashion design, in addition to the fine arts and crafts (Mulgan and Warpole 1992, Wynne 1992a). A wide conception of culture is embraced, and support is provided for new-technology cultural industries. Take, for example, Cardiff. This city claims to be the largest broadcast production centre outside London. Although much of this claim is based on its Welsh-language broadcasting, given that it largely services the 22 hours per week of programming by S4C, 5000 people are employed in media-related jobs in South Glamorgan. To encourage greater inter-linkages between firms in this sector, a directory of cultural industries has been produced listing the products and services which each provides (Wynne 1992c).

The principal method adopted by this approach, however, is to develop the infrastructure for cultural production through investment in studios, workshops, marketing and support agencies, usually in the form of cultural districts or quarters (Bassett 1993, Wynne 1992b). Although cultural quarters can contain a concentration of cultural production and/or consumption facilities and more often than not, combine both (Wynne 1992c), for those who adopt the production-oriented perspective, the emphasis is upon developing cultural quarters based on cultural production. Sheffield is an example of a city which to a greater extent than many others has pursued this type of cultural-quarter development, at least in the early stages of its cultural industry policy (see chapter 11).

It is fair to state, however, that the vast majority of localities pursue a strategy somewhere nearer the middle of this spectrum combining cultural production and consumption to varying degrees. Few, if any, places pursue either one or the other. It is the same, although to a lesser extent, so far as the goals of cultural industry development are concerned. Promotional strategies pursue 'flagship' projects to attract people and investment, for example through 'safe' forms of cultural provision in the form of 'high art', and more usually than not, do this implicitly and unintentionally. However, those who adopt the integrationist approach do so more explicitly and intentionally. Here, cultural provision is seen as a key to revitalizing public social life, reviving a sense of civic identity and shared belonging, creating a more inclusive and democratic public realm, and increasing expectations about what a locality has to offer (Montgomery 1990, Bianchini et al. 1988). Bologna and Rome are but two localities which have adapted this integrationist approach (Bianchini 1989, Bloomfield 1993), as are Dublin and Manchester (Montgomery 1995b, Urban Cultures Ltd 1992).

As McGuigan (1992) notes, moreover, this integrationist approach can adopt either a productionist trajectory emphasizing popular control of local cultural production such as through the promotion of community arts, ethnic minority cultures and art in social and culturally deprived neighbourhoods; or a consumptionist trajectory focusing upon the ways in which the products of mass entertainment can be creatively and critically appropriated by people, rather than passively consumed. To do this, wider public use of the arts is encouraged so as to open up traditional institutions such as museums and theatres to wider public use (Bassett 1993).

In sum, contemporary cultural industry strategies can adopt either a producer- or a consumer-oriented approach which may

be promotional or integrationist, although in practice most localities have a mixture of all of these to varying degrees. In a bid to
explain which strategy a locality will adopt, Griffiths (1995) has
argued that the approach will be a product of a mix of the following: the area's current and desired position in the regional and
international hierarchy; its recent and longer-term experiences of
economic and industrial restructuring; the underlying political culture of the locality; the political demands and priorities thrown up
by the area's changing social structure; the pattern of organization
of artistic/cultural and business elites; the opportunities made
possible by local administrative and governmental structures; and
the availability of government and other funding sources. Put
another way, he assumes that the approach adopted will be determined by the interplay between the structural factors regulating the
scope which a locality has to act, the particular characteristics of
the area and the nature of governance prevailing in that locality.
However, although this recognizes the structural and institutional
factors influencing strategy within areas, it under-emphasizes the
role of agency in the sense of key individuals who make cultural
policy happen, generate ideas or drive projects through to completion. Montgomery (1995b), however, highlights the importance of
project champions in the success of Temple Bar, Dublin. With this
in mind, the spatial variations in cultural industry strategies will
now be examined.

Spatial variations in cultural industry strategies

Unfortunately, there has been little structured research into how
these different types of cultural industry strategy are spatially
manifested. Instead, geographical inquiry has confined itself to
the extent to which localities have adopted and supported the
cultural industries. The most comprehensive analysis in this regard
is the work of Williams et al. (1995) on UK local authority policies
towards the cultural industries. The cultural industries, in this
instance, are taken to include the performing and visual arts, films,
broadcasting, photography, publishing, design and fashion, and the
heritage industry. A postal survey was undertaken of 198 district
councils in two standard regions: the 'North' (the northern region
and Yorkshire and Humberside) and the 'South' (the South-West
and the South-East), which received a 40 per cent response rate.
Responding authorities were also classified by whether they were

'more urban' or 'more rural' on the basis of their population size
and level of urbanization. The results of this survey are documented in Table 9.3.

Starting with the perceived role of the arts in local economic
policy, this study reveals that authorities view the arts as most
influential in improving an area's image (68 per cent) and attracting
tourists (62 per cent), but fewer consider them important in
attracting inward investment (41 per cent) and skilled workers
(31 per cent). Relatively few authorities, therefore, have grasped
the role of cultural industries in encouraging inward investment
and key workers, but more understand their role in promoting
cultural tourism and image enhancement.

At a regional level, although there is little difference in terms of
the perceived importance of cultural industries in attracting inward
investment and key workers, northern authorities have a greater
awareness of their role in inducing tourists and in improving the
image of the area, as witnessed in the rise of cultural tourism in

Table 9.3 Spatial variations in cultural industry strategies

	Region			
	North	*South*	*Urban*	*Rural*
Importance of arts in economic development				
Attracting inward investment	43	40	50	30
Attracting key workers	29	32	42	17
Attracting tourists	71	58	68	55
Improving the area's image	80	67	82	50
Direct support for arts				
Employment of an arts officer	32	31	45	15
% with arts policy	41	26	33	27
Per capita expenditure				
<£1.00	36	39	15	67
£1.00–£2.50	18	16	10	24
£2.50–£5.00	27	16	28	9
>£5.00	18	29	48	0
Specific activities promoted				
% promoting cultural festivals	64	49	72	44
% seeking to attract artists	23	14	23	9
% collaborating with arts companies	55	76	85	50

Source: Derived from Williams *et al.* (1995: Tables 2–6)

many northern cities (for example Bradford, Wigan) and their active ventures in place marketing (for example Glasgow, Manchester). Reflecting this, authorities in the North are more likely to have explicit arts policies than southern authorities. Urban areas, moreover, attach greater importance to the arts in economic development than rural areas, and this is reinforced by the budgets allocated to this sector. Urban authorities are also more likely to have explicit arts policies and a specific officer responsible for the arts than rural authorities. As Williams *et al.* (1995) thus conclude, this unlevel playing field in terms of policies for using the arts for economic ends and levels of funding reinforces current inequalities in cultural provision which means that urban authorities are better endowed with cultural provision than rural areas. Blau (1989) finds a similar tendency in the USA. For him, the solution is either for smaller localities to cooperate with neighbouring areas so as to plan cultural specialization and joint marketing, or for them to specialize in particular segments of the cultural industries market. Although there is little evidence of the former occurring in the UK, there are some signs of specialization by smaller rural areas, one example being Hay-on-Wye's annual literary festival in mid-Wales (see Tricker and Williams 1992).

However, it is not solely at an inter-regional and urban-rural level that there are inequalities. As Bianchini (1993b) highlights, there are also disparities at an intra-urban level. On the one hand, there are city centre/periphery tensions. Much cultural policy is increasingly targeted at city centres, especially flagship projects, resulting in the marginalization of the periphery. One solution is to create neighbourhood-based arts facilities, as in Hamburg and Bologna (Bianchini 1993b). On the other hand, there are tensions within existing cultural quarters, for example caused by gentrification, as displayed in New York's loft quarter (Zukin 1982) and Birmingham's Ladywood area (Loftman 1991). Here, protection can be offered to cultural producers who have been shown to suffer from low wage levels (British American Arts Association 1989, Wynne 1992c) so that they are not pushed out of such areas by rising prices.

In sum, although there has been increasing recognition of the role of cultural industries in economic development, this has been uneven in terms of the extent to which it has been recognized by authorities, how it has been incorporated and the magnitude of the investment. Although at an inter-regional level, peripheral regions, whether by choice or constraint, have been more adept at marketing their cultural resources than core regions (see, for example, Smales 1994), at an inter- and intra-urban, and urban-rural level, the result is more likely to have been spatial polarization. Unfortunately, however, and on account of lack of information, this analysis of spatial variations in cultural industry strategies leaves many questions unanswered. Is it, for example, the economic restructuring experiences of peripheral regions which have resulted in their being more active? Or is it the political cultures prevalent in such areas? Moreover, is activism in policy formulation any indicator of their success in cultural industry development? After all, although peripheral regions may be more adept at marketing their cultural resources, London is still way ahead in terms of external income generation from cultural industries (see chapter 14). These questions will need to be addressed in future research. It is not only at the issue of uneven development, however, that there are problems with the current nature of cultural industry policy so far as local economic development is concerned.

CULTURAL INDUSTRY POLICY: A CRITICAL REVIEW

Although the use of cultural policy in economic regeneration is increasingly an uncontested issue, several problems exist with the current way in which cultural industries are conceptualized as contributing to local economic development. These relate to how various external income-generating cultural industries are viewed, the relative importance attached to meeting local demand, the under-emphasis placed on local ownership and control of cultural industries and the need to strike a balance between the economic and social objectives of cultural industry policy.

First, there is a persistent tendency in the cultural industries literature to differentiate between the relative value of production-oriented and consumption-oriented cultural industries as generators of external income. Production-oriented cultural industries are seen as 'engines of growth' and high value-added industries offering the possibility of high-wage employment, whilst consumer-oriented cultural industries are conceptualized as relatively 'dependent' activities reliant upon other sectors of the local economy for their continued growth and providing only low-wage jobs. Bassett (1993), for example, asks how many cities can achieve success as major centres of cultural consumption, whilst Bianchini (1993b) argues that it can be risky for localities to rely on consumption-oriented strategies of cultural policy-led development. However, basic consumer service industries which generate external income by attracting cultural tourists and investment into a locality are as much part of the basic sector as those industries which export their product so as to generate external income. Neither is there any evidence as yet that producer-oriented cultural industries provide better-paid jobs than consumer-oriented industries. It is important to recognize, therefore, that both attract external income generation for localities, and one should not be seen as somehow superior to the other. The problem is perhaps that the production-oriented cultural industries are seen as closer to manufacturing-oriented activity, which is traditionally recognized as basic-sector activity, whilst consumption-oriented cultural industries are viewed as service-oriented industries, which are perceived as somehow inferior.

Second, much discourse in the cultural industries literature is dominated by a view that the external income-generating function of the cultural industries is their sole, or principal, contribution to economic development. As mentioned above, however, it is not external income generation *per se* but rather net income which is important for local economic growth. The role of cultural industries in preventing the leakage of income out of a locality is therefore equally important. There is, after all, little point generating external income if, at the same time, the locality leaks like a sieve, for example by its residents going elsewhere to meet their demand for cultural provision. This consideration, however, is frequently neglected. The result is that strategies focus upon meeting external demand, for instance through flagship projects, whilst there is a lack of emphasis on fulfilling local cultural demand. As Myerscough (1988) highlights, nevertheless, 90 per cent of expenditure of residents on local cultural provision would have gone on cultural provision outside the area if facilities had not been available locally.

The implication, therefore, is that there is also a requirement for cultural industries that retain income. These are not necessarily the same cultural industries as those which attract an external visitor or inward investment. The current focus upon developing prestigious high-profile 'flagship' projects so as to project images conducive to inward investment and attracting external visitors may be inappropriate for meeting local demand for cultural provision. Moreover, given that their development might result in the reduction of public subsidy for more community-oriented and smaller arts organizations (Bianchini 1993a) and more radical, risky projects and voices (Griffiths 1995), such outward-looking strategies may well increase, rather than reduce, the leakage of income. One example of 'good practice' in this regard is Temple Bar in Dublin, where policy attempts to strike a balance between these two types of demand (Montgomery 1995b).

A third problem with current cultural industry strategies relates to the extent to which they facilitate local control and ownership of cultural production and consumption. As in other industrial sectors, the tendency in cultural provision is towards globalization and the concentration of capital. The result is an increasing homogenization of cultural provision (Boyer 1992, Commission of the European Communities 1990) and an oligopolistic industrial structure. Take the music industry, for instance. In 1992, five major global corporations dominated the industry: EMI, Sony, Polygram, BMG and Warner Brothers/Time Warner. These captured 73 per cent of sales in the USA, 60 per cent in Japan, 90 per cent in Germany, 73 per cent in the UK and 87 per cent in France (Monopolies and Mergers Commission 1994). Smaller record companies, moreover, tend to act almost as research and development centres, spotting and cultivating new trends and styles, and are often affiliated to the five majors (Leyshon *et al.* 1995). Similarly, vertical and horizontal integration in the cultural industries means that many other forms of cultural provision are owned by large companies, such as Bass, Granada and Rank International, ranked the thirty-first, forty-sixth and fifty-fifth largest companies in the UK by market capitalization (*Sunday Times*, 19 November 1995). The resulting problem is that cultural production and consumption are increasingly externally owned and controlled. The development of flagship projects often serves merely to reinforce this trend. Cultural industry strategies, therefore, need to develop policies to promote local ownership of cultural production and consumption facilities. If the intention is to maximize net income for the locality, then much greater emphasis will need to be given to considering ways in which locally owned indigenous enterprises can be supported and developed.

What is thus required is greater focus upon promoting local entrepreneurship, local ownership and the meeting of local demand so as to create denser local inter-linkages at the interface of cultural production and consumption. The result will be increased local employment multipliers since neither profits nor spending will flow out of the local economy to the same extent as at present. Up to now, the problem has perhaps been that cultural industry strategies have not always clarified either their aims or whom they wish to be the major beneficiaries of such policies. Indeed, it is perhaps the case that there is much to learn from the community cultural planning approach which has been widely adopted in the USA. Dreeszen (1994: 19) defines this as 'a structured community-wide, public/private process, that engages the members of the community in communications to identify their community's arts and cultural resources, needs and opportunities, and to plan actions and secure resources to address priority needs'.

The final problem with cultural industry strategies is their focus upon economic objectives at the expense of social objectives. Cultural industry strategies, however, are not only tools for maximizing net income. They are also a potential mechanism for enabling greater social integration in localities and for reviving civil society. To this end, Bianchini (1993b) has argued for the adoption of a holistic 'cultural planning' approach which has a much broader view of cultural industry development than the current economic development approaches. Rather than viewing this sector simply as an economic tool, this approach sees it as a response to the crisis in the texture, the richness and the meaningful content of daily life. The cultural depletion which it seeks to redress includes the fear of walking on the streets, the deep mistrust of strangers, the demise of civil society and social cohesion, the decline of conviviality, the homogeneity of city centres, the loss of public spaces and the poverty of the public realm (Bianchini 1993b, Comedia 1991, Griffiths 1993, Landry and Bianchini 1995). The aim here, therefore, is to recreate public spaces, opportunities for interaction and greater tolerance of difference. This is not incompatible with economic development. Rather, it is precisely about improving the quality of life in the public realm which is so strikingly missing in the vast majority of localities and which can prove attractive to investment.

Whether or not this more locally oriented holistic approach is pursued, it is certain that the cultural industries can no longer be seen as dependent activities. Instead, this chapter has revealed that they are both generators of external income and a valuable resource upon which the future quality of life in many localities depends. It is time, therefore, that localities started to reinvest resources in a consumer services sector which can provide them with a much-needed boost in the current era of inter-locality competition.

Part III

LOCALITY CASE STUDIES

10

GLASGOW
A city of culture?

INTRODUCTION

Glasgow, with a population of 662000 in 1991, is the largest city in Scotland. As one of the first cities to witness deindustrialization and with a distinct lack of success in attracting producer services, Glasgow had little other option open to it than to look elsewhere for a sector which could help it climb out of the mire. In these circumstances, Glasgow has attempted to establish itself as a 'city of culture'. The aim of this chapter, therefore, is to use this city to evaluate the role that cultural industries can play in local economic development.

To commence, however, and given that approaches towards local economic development do not arise in a vacuum but are the result of the interplay between global trends, local circumstances and institutional policy regimes, the nature and extent of the problems confronting Glasgow will be charted. With this contextualization in hand, the 'city of culture' approach adopted in this city, with its reliance on the cultural industries sector, will be analysed and the economic impacts of this economic development strategy evaluated. Finally, and arising out of this analysis of the approach embraced and its impacts, conclusions will be drawn about the changes required to make this strategy more effective. First, however, a brief historical excursus is required to provide the background of the decision taken to adopt an approach towards rejuvenation centred on the cultural industries sector. This will reveal how consumer services, at the bottom of the sectoral hierarchy of basic activities, were adopted only because no other options appeared to be either available or realistic.

THE DEINDUSTRIALIZATION OF GLASGOW

Since it was one of the first European settlements to industrialize, it was perhaps inevitable that the problems associated with deindustrialization should be felt first and most acutely in Glasgow and its surrounding area of west central Scotland (Lever 1992, Pacione 1995). This area developed as an industrial centre in the seventeenth and eighteenth centuries, processing raw materials such as sugar, cotton and tobacco from the trans-Atlantic trade. Its most rapid period of development, however, was the latter part of the nineteenth century, when its steel, engineering and shipbuilding industries expanded. Shipbuilding, in particular, characterized the city at this time. By 1903, there were 39 shipyards in Glasgow with a workforce of 100000 launching a vessel per day, a total which represented one-third of all British and one-fifth of total world production (Lever 1992). Indeed, it has been estimated that in 1900, almost one-half of the world's sea-going vessels had been built on the Clyde (Keating 1988). The latter part of the nineteenth century was thus the apex of Glasgow's industrial power as 'the second city of the Empire'. Indeed, its population grew from 77000 in 1801 to ten times that in 1911, reaching a peak of over 1 million in 1931 (McCrone 1991).

As MacInnes (1995: 76) notes, however, 'If the rise of Glasgow's industrial power was fast, so too was its fall.'[1] The interruption to world trade patterns caused by the First World War and the depression of the early 1930s was to mark the start of its decline. For example, by 1933, the Clyde launched just 56000 tons of ships: 7 per cent of the 1913 level. Indeed, seven out of ten Scottish shipbuilding workers were jobless while overall, one in three men in Glasgow were recorded as unemployed (Keating 1988). However, although Glasgow's industrial problems were clearly visible in the inter-war period, these became temporarily masked by the Second World War and the boom in international trade following 1945 (Lever 1992, Pacione 1995). Rearmament and then replacement to some extent revived the city's traditional industries. Some 400000 people continued to be employed in the conurbation's manufacturing industries, and in the 1950s the Clyde launched around 400000 tons of ships each year. For those seeking the origins of Glasgow's decline, nevertheless, the signs were clearly visible. This unchanging tonnage represented a steadily declining share of world supply (Foster and Woolfson 1986, Randall 1980).

Moreover, much of the post-war reindustrialization of Glasgow was based on inward investment and thus suffered from the inherent problem of external control in a world of increasingly mobile capital (MacInnes 1995).[2] Despite the rising prevalence of these warnings signs for the stability of the Glasgow economy, the 1950s were a decade in which few paid much heed to their potential implications. Indeed, and as Table 10.1 shows, there seemed little reason for concern. Between 1951 and 1961, not only did the absolute number of jobs in the city increase, but so too did the absolute and relative number of jobs in manufacturing.

From the 1960s onwards, however, the situation rapidly changed. Glasgow started to witness both an absolute and a relative decline in its manufacturing employment. As Table 10.1 illustrates, total employment in the city declined by over a third between 1961 and 1991, and the bulk of these losses were in the manufacturing sector. Indeed, three-quarters (74.5 per cent) of manufacturing jobs were lost. The seeds of decline, so clearly visible to those who wished to see them in earlier decades, had germinated.[3] Glasgow's over-reliance on mature and declining manufacturing sectors (for example shipbuilding and mechanical engineering), a lack of investment in new technologies and the emergence from abroad of strong competition for its markets (Lever 1992, MacInnes 1995, Pacione 1995) had combined to produce a rapid decline in manufacturing employment. This was further compounded by the general deindustrialization of the UK economy as manufacturing shifted to other, lower-cost nations, by the intra-national shifts in such economic activity towards the South-East and the shift in manufacturing within the conurbation away from the city and

Table 10.1 Employment change in manufacturing and services in Glasgow, 1951–91

	Total employment (000s)	Manufacturing total jobs (000s)	% of total	Services total jobs (000s)	% of total
1951	521	265	50.8	255	48.9
1961	540	279	51.7	260	48.1
1971	466	192	41.2	273	58.6
1981	399	135	33.8	263	65.9
1991	337	71	21.1	263	77.3

Source: Census of Employment 1991

towards greenfield sites and industrial estates beyond its boundaries in the smaller towns and new towns around the conurbation
(Lever 1992, MacInnes 1995, Pacione 1995).

As in many other cities, moreover, the growth of the service
sector in Glasgow was insufficient to compensate for this decline
in manufacturing jobs.[4] Indeed, in Glasgow, the absolute number
of service-sector jobs hardly changed between 1961 and 1991,
even though the percentage employed in services continued to
rise on account of the much quicker rate of decline in manufacturing jobs. A major reason for this lack of growth in service-sector
jobs is the depopulation of Glasgow city. Between 1961 and 1991,
for instance, the population of the city declined from 1.1 million to
662000, virtually mirroring the rate of urban growth which had
taken place a century earlier (Paddison 1993). Other reasons for
this lack of service-job growth are the movement of services out of
the city, with such jobs increasing in the outer conurbation faster
than in Britain in general (MacInness 1995), and its apparent
inability to attract producer services to the same extent as Edinburgh, its powerful neighbour, which remained the centre for
financial services in Scotland.

By 1991, therefore, and as Table 10.2 reveals, Glasgow had
become a city relatively more dependent on service employment
than the British economy as a whole. Indeed, this dependency
became exaggerated during the 1980s. This was not due to the
growth in service-sector jobs, which grew only by 1.1 per cent
between 1981 and 1991 compared with a national increase of 17.1
per cent. Rather, it is due to the rate of manufacturing decline. The

Table 10.2 Sectoral distribution of employees in employment in Glasgow,
1991

	Number (000s)	% of employees	Location quotient
Primary, manufacturing and construction	76.3	22.7	0.78
All services	260.3	77.3	1.09
Consumer services	145.3	43.2	1.08
Producer services	35.0	10.4	0.96
Mixed producer/consumer services	53.5	15.9	1.16
Public services	26.4	7.8	1.24

Source: Census of Employment 1991

service sector in Glasgow, therefore, and contrary to some of the images portrayed by the city marketeers, is not rapidly expanding, although it is developing more rapidly than the other sectors of the economy. As Pacione (1995) highlights, almost every SIC group showed a decline in employment between 1981 and 1991, with only insurance/banking (+23 per cent), public administration (+30 per cent) and other services (+10 per cent) showing a rise.

By 1991, therefore, the economic structure of Glasgow was one in which the primary, manufacturing and construction industries as well as producer services were under-represented relative to the British economy, whilst consumer services, mixed producer/consumer services and public services were over-represented. This, however, and as highlighted above, is more due to the decline in its manufacturing sector and the relatively slow growth of producer services than any massive increase in employment in its consumption-oriented industries. This analysis reveals, therefore, that Glasgow is shifting not so much from an industrial metropolis to a city of consumption as from a manufacturing to a non-manufacturing economy. Services have become more important by default as the manufacturing sector has spiralled into terminal decline.

The outcome of this economic restructuring process has been severe socio-economic problems.[5] In 1981, male unemployment was 19.4 per cent, and this had risen to 25.7 per cent by 1991, whilst female unemployment rose from 10.5 per cent to 15.3 per cent over the same period (Pacione 1993). Some 29.6 per cent of the city's population, moreover, are dependent upon state income support (Paddison 1993). These socio-economic problems are unevenly distributed. As in many other large urban areas, a socio-spatial apartheid is in operation, albeit with a dynamic which has shifted over time because of public-policy interventions (Donnison and Middleton 1987, Pacione 1995). With the clearance and redevelopment of the deprived inner-city areas, the major areas of deprivation in Glasgow have moved outwards onto the peripheral estates such as Drumchapel, Castlemilk, Easterhouse and Pollok and into inner suburban areas such as Possilpark and Haghill. On the disadvantaged council estate of Easterhouse, for example, infant mortality rates are five times higher than those in the nearby middle-class suburb of Bishopbriggs (Pacione 1993). Given these socio-economic problems and the rapid rate of deindustrialization, action was required.

GLASGOW'S 'CITY OF CULTURE' APPROACH

In 1981, confronted with both a deindustrializing city and severe socio-economic problems, Glasgow District Council (GDC) established an Economic Development and Employment Committee with the sole remit of generating employment. To do this, it launched a number of conventional initiatives such as local retraining schemes, upgrading local office and industrial land and premises, and providing financial help to local businesses. The central policy tool, however, was changing the image of the city so as to attract inward investment (Paddison 1993). As Taylor (1990: 2) puts it, 'Glasgow was seen as the City of mean streets and mean people, razor gangs, the Gorbals slums, of smoke, grime, of drunks, impenetrable accents and communists.' A new image, it was hoped, would encourage key decision-makers to invest in Glasgow and might attract key personnel by persuading them to move to Glasgow for a better quality of life (Boyle and Hughes 1991).

As the 1980s progressed, however, it became increasingly apparent to the GDC that this strategy was not working and that tourist expenditure *alone* (stimulated by the new image) might be significant enough to generate a substantial number of service-sector jobs (Boyle and Hughes 1991). Indeed, for many, this was the only avenue left open for them to follow. As Boyle and Hughes (1994) document in some detail, this conclusion was reached via several consultancy reports. Reports by both McKinsey & Co. and Gordon Cullen recommended that the city centre could be rejuvenated using the cultural industries in particular and consumer services more generally. This was further reinforced by Myerscough (1988), who was commissioned by the Glasgow District Council to examine the economic importance of the arts to Glasgow. The outcome, as Paddison (1993) puts it, was that there was a shift from indirect marketing to import producers to more targeted forms of marketing to import consumers 'focused around the arts and culture as a means of increasing tourism and of conveying the notion of Glasgow as a post-industrial city' (Paddison 1993: 347). Glasgow, the 'city of culture', was born.

In order to achieve this, and as Lim (1993: 592) has highlighted, 'Glasgow's approach has been a mixture of measured incrementalism and opportunistic adventurism.' On the one hand, and since the early 1980s, there has been the upgrading of its cultural and

152

arts infrastructure through substantial public investment. This commenced in 1983 with the opening of the Burrell Collection in Pollock Country Park at a cost of £23 million. Following this, the new Royal Concert Hall opened in 1990, the £36 million Scottish Exhibition Centre was constructed, a new home was built for the Royal Scottish Academy of Music and Drama, and in 1996, the £7 million Gallery of Modern Art was opened. In 1999, moreover, the £40 million National Gallery of Scottish Art will open. Alongside these cultural developments, a number of shopping centres were also erected in the city using private capital, including Princes Square (which opened in 1987), the Forge (1988), St Enochs (1989) and the Italian Centre (1990). In late 1995, moreover, work began on the 600000-square-foot Buchanan Galleries shopping centre in the city centre. On completion in spring 1999, the £150 million project will be Scotland's largest city-centre shopping development. There has also been considerable private-sector investment in housing, offices and hotels in the city centre. Following the opening of a major international hotel at the Exhibition Centre in 1989, Hilton International built the city's first full five-star hotel at a cost of £27 million with 20 storeys, 319 bedrooms, a specialist conference centre, 11 boardrooms and a ballroom for 1000 people. Indeed, between 1985 and 1991 in the city centre, about 150 developments with an aggregate value of around £1.8 billion were completed (McCrone 1991).

To utilize these infrastructure developments in a positive way to generate external income and market the city, Glasgow has adopted an event-led strategy towards tourism promotion. In Glasgow, and unlike Sheffield (see chapter 11), such a strategy is not new. In 1888 Glasgow had been the venue for the first of four exhibitions which promoted industry, art, history and science, following Queen Victoria's Jubilee. In 1938, moreover, the last and largest Empire Exhibition was staged in Bellahouston Park; it attracted 13.5 million visitors (Heeley and Pearlman 1988). The start of the contemporary event-led strategy, nevertheless, was in 1982, when GDC launched Mayfest, an annual arts festival. Conscious of the lack of other hallmark events in the cultural calendar to attract tourists, the city was then led by its 'opportunistic adventurism' to successfully bid to host the National Garden Festival in 1988, then to become the 1990 European City of Culture and the 1999 City of Architecture.

Take, for example, the 1988 National Garden Festival. In 1983,

GDC submitted a proposal to stage the third national garden festival on the site of the redundant Princes Dock (120 acres) on the south bank of the Clyde adjoining the area of Govan. German in origin, and used extensively in the post-war period to repair war damage, the garden festivals are intended to reclaim derelict land and/or refurbish existing parkland through major exhibitions of plants. The exhibitions are then removed, leaving the upgraded land for future development. In Glasgow, and in a bid to heighten its profile as a tourist destination using an event-led approach, the preliminary estimate was that the capital costs (including site acquisition, reclamation and festival development) of the garden festival would be £23 million, with a further £3.5 million operating costs. Visitor income was projected at £8.5 million and the value of the residual assets at £9.5 million. The net cost of the festival was thus put at £8.5 million. In fact, the festival cost £41 million to stage and the 4.25 million visitors generated £23 million of revenue, giving a net cost to public funds of £19 million (Heeley and Pearlman 1988). The land on which the festival took place, moreover, has remained largely undeveloped, and there is a continuing debate about its future use.

Without going into the success or otherwise of this event, it is evident from an examination of both the facilities constructed in Glasgow and the events hosted that its 'city of culture' approach is consumption-oriented in the sense that the intention is to attract tourists into the locality in order to benefit from their spending. There is little if any evidence of production-oriented cultural industry developments. It also seems from the above that the goal of such cultural industry development is essentially 'promotional' rather than 'integrationist', given the emphasis put on flagship projects so as to import consumers (see chapter 9). However, such 'broad-brush' sketches of a city's approach do not always reveal the reality. Here, therefore, and using the example of the 1990 European City of Culture festival, a finer-grained analysis of Glasgow's approach towards the cultural industries is undertaken. The aim, in so doing, is to show that great care must be taken before pigeon-holing the approach of individual cities towards cultural industries development from overarching sketches. An analysis of this hallmark event will reveal that despite Glasgow being widely perceived as pursuing primarily a consumption-oriented strategy and promotional goals, there have been production-oriented elements in its strategy and it has managed to retain to some degree the goal of integrationism so far as cultural provision is concerned. The reality of its cultural industries strategy, therefore, is more complex than it at first appears. It is to this that attention now turns.

The European City of Culture 1990 hallmark event

In June 1985, the Ministers responsible for cultural affairs within the European Community established the 'European City of Culture' designation with the aim of allowing the chosen city to open up to the European public particular aspects of its cultural heritage and to enable a number of cultural contributions from other countries to be concentrated in the designated city. Following Athens (1985), Florence (1986), Amsterdam (1987), Berlin (1988) and Paris (1989), Glasgow became Europe's sixth City of Culture in 1990.[6]

Once the bid was accepted, the Labour-controlled GDC assumed responsibility for financing and organizing the event. A total of £15 million was added to the normal £20 million which GDC spent on cultural activities (Boyle and Hughes 1991). Other contributions came from Strathclyde Regional Council (£12 million), private sponsors (£6 million), the Office of Arts and Libraries (£0.5 million) and the European Community (£0.5 million). With a total budget of £54 million, over 3800 events were eventually held, at least one taking place on each of the 365 days of the year (Boyle and Hughes 1994). Of the £15 million special fund donated by the GDC, over £8 million went on one-off public relations-oriented events, including £4.6 million on the 'Glasgow's Glasgow' exhibition tracing the city's 800-year history (a major loss-maker), £1 million on the visit of the Bolshoi opera, £1 million on the Big Day, a rock and pop festival, and £650000 on concerts by Frank Sinatra and by Pavarotti. A further £4 million was spent on direct marketing. Saatchi and Saatchi, the international marketing consultants, were employed to market the event to south-east England, for instance.

The focus, therefore, was upon spectacular consumption-oriented events with promotional objectives. Indeed, this is reflected in the original bid document to host the 1990 European City of Culture event. In setting out its objectives, GDC explicitly emphasized the importance to the city of developing its cultural tourism sector as a tool for economic regeneration (Glasgow

District Council 1987). From the outset, therefore, a consumption-oriented promotional approach was envisaged. As Booth and Boyle (1993: 44) state, 'it is clear that Glasgow's adaptation of 1990 European City of Culture was primarily concerned with the use of culture for urban marketing and tourist promotion'. Superficially, the European City of Culture hallmark event could thus be interpreted as reinforcing the view that Glasgow's 'city of culture' approach is essentially promotional-oriented and consumption-focused. However, to paint such a 'broad-brush' picture of this hallmark event would be to do an injustice to both the rationale underlying it and the way in which it operated in practice.

As with much of Glasgow's 'city of culture' approach, although the main thrust of this event was promotional and consumption-oriented, elements of both integrationist goals and a production-oriented strategy can be identified. The European City of Culture event included the promotion of community arts, ethnic minority cultures and art in socially and culturally deprived neighbourhoods. Theatre groups and arts centres from peripheral housing estates organized plays and photographic exhibitions. Artists in residence were located in areas such as the Gorbals, whilst the Jewish and Irish communities organized extended programmes and local communities were trained in the art of commissioning sculptures with special local significance. Hence, under the predominant discourse of consumption-oriented events for cultural tourists lay a dense web of production-oriented events, conceived and organized by community groups with integrationist objectives in mind.

As Booth and Boyle (1993: 45) conclude about the 1990 event, 'there is no evidence . . . that the spectacle was used as a reward for the middle class. The range of events was not divided by class. The scale of the community programme suggests that resources were allocated widely throughout the city.' Indeed, the Labour group on GDC continually asserted that the event was a celebration of local culture, not solely high art, and that it was not as elitist as it had been in other cities in the previous years. Reinforcing this assertion, Booth and Boyle (1993: 35) state of the year, 'The emphasis was upon endeavour rather than high or low art. The implementation of the programme had two strands: to support local artists and groups on their own terms and organise a broad cultural programme that would raise Glasgow's profile in Europe and beyond.' Indeed, Booth and Boyle (1993: 45) go so far as to assert that it was used 'as the mechanism for urban unification', although whether it was successful or not in this task is open to question.

It is doubtless the case, therefore, that the City of Culture event possessed both integrationist goals and a production-oriented trajectory. The only disagreement is over the extent to which these were prevalent and whether greater emphasis should have been placed on them. For some in Glasgow at the time, neither the integrationist goals nor the productionist trajectory was sufficiently to the fore to make the event acceptable. One particularly prominent group in this regard was 'Workers' City', a loose collection of Left-orientated local artists, celebrities and other local worthies. Although not representative of the Glasgow population, this was the loudest and most articulate voice opposing the event. The thrust of the Workers' City's critique of the European City of Culture year is set out in their two manifesto publications (McLay 1988, 1990). These have been assessed in depth elsewhere (Boyle and Hughes 1991, 1994). The important point to note here is that they believed that the event was oriented too much to the tastes of middle-class and tourist consumers, arguing that the emphasis on elitist 'yuppie' cultural activities not only undermined the city's natural working-class cultural heritage but displaced working-class community events (Boyle and Hughes 1991). They also questioned the economic benefits of such an approach to regeneration, arguing that it would only profit mobile capital and benefit the affluent in the city. However, and as Booth and Boyle (1993: 39) assert,

> the criticism of Workers' City needs to be seen in the context of an extensive community events programme that attracted widespread support during 1990 . . . a range of more than five hundred exhibitions, local gala days and theatrical events brought the year of Culture closer to the public, especially into the peripheral public sector housing schemes that lacked an indigenous, well-established, arts community.

Indeed, it is telling that Workers' City has broken up and that many of its protagonists now praise the developments that have taken place (Arlidge 1996). The objections of Workers' City to the 1990 event, however, related not solely to the strategy and goals adopted but also to the economic impacts of this approach. To examine whether this is and was a valid criticism, the next section evaluates both the economic impacts of this hallmark event and the wider overall economic impacts of using the 'city of culture' approach to regenerate Glasgow.

ECONOMIC IMPACTS OF THE 'CITY OF CULTURE' APPROACH

In assessing the role of the 'city of culture' approach in regenerating Glasgow, it has to be remembered that this city has rapidly lost both population and jobs and that it has had a tarnished image.[7] Given the deindustrialization processes at work in this local economy, and the traditional competitive advantage of Edinburgh so far as producer services are concerned, such consumer services were perhaps one of the few options open to the city so far as economic development was concerned. Cultural industries in particular, and consumer services more generally, thus became the lead sector in regeneration efforts in Glasgow rather than merely a sector supporting investment in other sectors because of its entrenched economic position. How successful, therefore, has this approach been in rejuvenating Glasgow?

On the issue of external income generation, there is little doubt that the number of visitors to Glasgow has significantly increased. Before the promotional campaign based on the 'Glasgow's Miles Better' slogan began, the standard local joke was that if a tourist was in Glasgow, it was because s/he was lost. In 1982, nevertheless, as the campaign to attract tourists began, there were 700000 visitors generating £68 million of expenditure (McCrone 1991). During 1990, meanwhile, there were 1.75 million (Scottish Tourist Board 1995a and b). Consequently, and if Myerscough's (1988) estimate is used that an extra 50000 tourist trips to Glasgow provide 338 more jobs in the city, employment resulting from tourism can be calculated to have increased from 4732 in 1982 to 11830 in 1990. Whether this is an accurate portrayal of the situation or not, the evidence from the scale of the developments taking place in Glasgow over this period to accommodate its growing share of the tourist market is that there was rapid growth. During the late 1980s, for example, twelve major hotel projects were announced based on a private-sector investment of £116 million in the city (Booth and Boyle 1993).

The cultural industries, moreover, do appear to have played a key role in generating this increase in visitor numbers. In 1985, the arts in Glasgow were a £204 million industry employing 14735 people

(2.25 per cent of the working population) either directly or indirectly, more people than built ships on the Clyde (Myerscough 1988). By 1989, Booth and Boyle (1993) estimate that this had increased to 25000 jobs. This 69 per cent growth in arts employment in just four years is certainly a significant increase in a city which, it has to be remembered, was witnessing a continuing reduction in its employment base throughout this decade. This growth, furthermore, was before the 1990 European City of Culture event had even taken place.

To assess the impacts of this hallmark event, meanwhile, is relatively easy compared with other such events, principally because this is one of the few for which a retrospective analysis has been undertaken. Myerscough's (1991) study conducted immediately after the event finds that this £54 million investment brought a 'net return' of £10.3–£14.1 million and created 5350–5580 'person years" worth of jobs. Gross public expenditure per job was £7286, which compares favourably with estimates for other initiatives (see Foley 1991). Hotel and guest house occupancy rates, for example, witnessed a substantial increase from 1989, with bedspace occupancy rising by 39 per cent and room occupancy growing by 35 per cent (Scottish Tourist Board 1990, in Booth and Boyle 1993).

The wider longer-term benefits of this hallmark event, however, have yet to be calculated and, as outlined in the discussion on hallmark events in chapter 7, it is perhaps here that the major impacts are to be found. If the pre-1990 and post-1990 numbers of tourists are compared, for example, a significant increase in overseas tourist numbers and expenditure can be seen. Between 1983 and 1989, the average annual number of overseas tourists was 264000, spending an average of £71 million per annum in 1994 prices (Scottish Tourist Board 1995a). In the period 1991–4, however, and as shown in Table 10.3, 455000 on average visited per annum (a rise of 42 per cent on the pre-1990 level), spending an average of £112.25 million per annum (an increase of 37 per cent). Given that overseas visitors and expenditure in Scotland as a whole increased by only 24 per cent and 34 per cent respectively during the same period, Glasgow's growth rate in attracting foreign tourists was higher than that of Scotland as a whole. However, whether this is due to the City of Culture event or whether it is due to the general increase in urban tourism cannot be known.

Further evidence of the nature of the impacts of Glasgow's

Table 10.3 Tourism visits and expenditure in Greater Glasgow, 1990–4

	British tourists		Foreign tourists		Total	
	Trips (m)	Expenditure (£m 1994 prices)	Trips (000s)	Expenditure (£m 1994 prices)	Trips (m)	Expenditure (£m 1994 prices)
1990	1.3	150	450	130	1.75	280
1991	0.8	99	430	89	1.23	188
1992	0.7	91	460	106	1.16	197
1993	0.8	97	440	108	1.20	205
1994	0.8	110	490	146	1.29	256

Source: Derived from Scottish Tourist Board (1995a and b)

event-led strategy is provided by a survey of white-collar house-holds in south-east England at the start and end of 1990. This recorded substantial increases in those feeling that Glasgow was 'rapidly changing for the better' (rising from 34 to 49 per cent) and a decline in the share considering the city to be 'rough and depressing' (reported in Paddison 1993: 347). However, the European City of Culture event did not affect the desirability of Glasgow so far as actually moving there was concerned: only 10 per cent 'would be happy to live and work' in Glasgow, a proportion which remained unaffected by the event (Paddison 1993). Paddison (1993) explains this not as a failure of the hallmark event but as a result of the fact that for an individual, the image of a city is compartmentalized. Although hallmark events may alter specific aspects of a city's image in an individual's mind, which in turn may alter the decision to visit the city, such events may leave the overall image of the city relatively unaffected, especially its more negative aspects. If this is correct, it has major implications for using an event-led strategy to reconstruct the image of a city for potential investors since it implies that it will not alter the overall perception of the 'quality of life' to be had in such localities and entice them to relocate. Hence, such marketing may be useful for promoting cultural tourism but not inward investment or the relocation of the 'new service class'.

Even in the realm of attracting tourists, however, Paddison (1993) argues that such hallmark events have only transient effects. This is supported by the tracking surveys of the 1990 European City of Culture event, where even by the beginning of the following year perceptions of the city, both in general and as a cultural centre, were reverting slightly and significantly fewer saw it as an 'exciting place to visit' (Myerscough 1991: 209). So, not only may event-led strategies have little impact on inward investment decisions, but they may also have a limited life so far as cultural tourism is concerned. The implication, therefore, is that if an event-led strategy is pursued, then a city will have to seek new hallmark events continually to retain its share of the tourist market and remain a high-profile place to visit. This short-term impact of hallmark events on tourist behaviour, however, is but one of the problems in the current 'city of culture' approach towards regeneration.

CONCLUSIONS

One of the principal problems confronting anybody wishing to evaluate the impacts of Glasgow's 'city of culture' approach towards local economic development is the lack of empirical data. Indeed, although much has been written on the economic problems confronting Glasgow and the political issues surrounding the 1990 European City of Culture event, there is remarkably little evidence on the impacts of the cultural industries on the regeneration of Glasgow in the contemporary period. What little evidence does exist, nevertheless, and as has been revealed, demonstrates that the cultural industries are growing and that tourism is playing a larger part in the Glasgow economy.

Here, however, the intention is to conclude by highlighting some of the ways in which Glasgow's 'city of culture' approach might be reoriented to increase its impacts on both economic and social development. To do this, it will be argued that the relative importance attached to meeting local demand and preventing leakage of income needs to be heightened, that greater emphasis needs to be placed on local ownership and control and that there is a need to strike a balance between the economic and social objectives of cultural policy.

Until now, and as shown above, the 'city of culture' approach has principally focused upon promoting the city so as to attract external income via cultural tourists. The resulting problem is that relatively little importance has been attached to meeting local demand and preventing income from leaking out of Glasgow. This, as has been argued throughout this book, has important economic implications for net income. Take, for example, the focus in Glasgow's cultural policy upon consumption-oriented rather than production-oriented cultural industries. This has resulted in local talent leaving the city since the infrastructure for retaining them is not in place. A new strategy is thus required to nurture, train and support talent locally and provide them with accommodation in the city (Booth and Boyle 1993). Similarly, there is no policy of seeking vertical integration in the cultural industries in the sense that the higher value-added stages of the cultural production process are not being put in place. The consequence is that 'whilst new goods may be conceived in Glasgow at the moment, facilities elsewhere are used to create such goods' (Booth and Boyle 1993: 43).

There is also a need for much greater attention to be paid to meeting local demand rather than external desires. As Myerscough (1988) has highlighted, 90 per cent of the expenditure on cultural provision by residents is 'ring-fenced' by people specifically for this purpose and will be spent outside the locality if facilities are not available within localities. The cultural provision required to meet this demand, however, is not necessarily the same as that which attracts an external visitor or inward investment. The current focus upon developing prestigious high-profile 'flagship' events and facilities so as to project images conducive to inward investment and cultural tourists may not only be inappropriate for meeting local demand for cultural provision but may also increase the leakage of income. However, little, if any, research has been undertaken into the nature and proportion of local cultural expenditure which takes place outside Glasgow, and neither has there been any investigation of whether the current mix of cultural facilities is sufficient to match the local demand for cultural goods and services. The implication, therefore, is that much greater attention is required to retaining local income.

A further issue is that the current strategy places very little emphasis on local control and ownership of cultural consumption and production. The resulting problem is not only that facilities are increasingly externally owned (for example the hotels, retail outlets, venues) but also that this is producing a homogenization of cultural provision resulting in Glasgow's becoming like everywhere else rather than celebrating and advancing its own cultural heritage. To produce a differentiated cultural product for tourists (as well as to enhance civic pride) and to keep local expenditure on cultural provision in the city, policies are required to promote locally owned indigenous cultural enterprises.

The final problem is that the focus of cultural policy has perhaps been upon economic objectives at the expense of social objectives. Cultural policy, that is, is not only a tool for economic regeneration. It is also a vehicle for facilitating greater social cohesion in localities and for reviving civil society. Glasgow could thus learn much from the 'cultural planning' approach highlighted in chapter 9 which transcends the narrowly defined perspective which views cultural industry development solely as an economic-regeneration tool and instead perceives it as a response to the crisis of daily life by recreating public spaces, opportunities for interaction and the tolerance of difference (for example Landry and Bianchini 1995). Neither is such an holistic approach incompatible with economic regeneration. Instead, it is precisely the enhancement of the quality of life in the public sphere which is missing from so many localities and which can prove attractive to investment. It is also a means of answering the criticism that the widespread areas of deprivation lying beyond 'culture city' are being by-passed by an urban strategy which is primarily concerned with increasing Glasgow's attractiveness as a locale for investment and tourism (Damer 1990, Mooney 1994a and b).

None of these problems mean, however, that this 'city of culture' approach should not have been adopted. Indeed, in the context of the broader structural processes taking place, it is probably correct that this was one of the few options open to Glasgow. Rather, what is being highlighted here is the future refinements in strategy required for this 'city of culture' approach to become more effective as a tool for both economic and social development. Whether or not these modifications are adopted, it is certain that Glasgow is an example of a city which has recognized the value of cultural industries to its economic regeneration and has put them at the fore of its strategy for rejuvenation. The view that such consumer services are dependent activities has been actively and prominently replaced with an acceptance that they can act both as a motor of economic development and as a catalyst for improving the quality of life. For further progress to be made, however, greater emphasis must now be placed on nurturing and retaining the city's indigenous cultural production and consumption, for it is upon this that its future cultural and economic vitality and viability ultimately depend.

11

BEYOND STEEL CITY

Consumer services and economic development in Sheffield

INTRODUCTION

As Britain's fifth largest city, with a population of 527400 in 1991, Sheffield conjures up several contrasting visions in the popular imagination. There is first its old image as 'steel city', a grim dour northern place where even the sparrows cough and second, its new image being promoted by Sheffield's development agencies as a place of culture, sport, leisure and pleasure. The aim of this chapter is to evaluate critically whether this transformation has occurred in reality. First, the extent to which Sheffield is shifting from being a manufacturing city to a locality based on consumer services will be examined. Following this, an exploration is undertaken of the magnitude and character of various sub-sectors of consumer services in Sheffield, starting with tourism and moving through sport, cultural industries, retailing and higher education to medical services. In the process, it will be revealed that the conventional perception of consumer services as 'parasitic' activities, supposedly reliant upon other, wealth-creating, sectors of the economy (manufacturing and business services) for their vitality and viability, is far from the reality in contemporary Sheffield. Since the mid-1980s, consumer services will be shown to have played a dominant role in Sheffield's economic revitalization.

THE DEMISE OF STEEL CITY

The old Sheffield was a belching steel-producing city. By 1850, it was supplying 90 per cent of all British and 50 per cent of European steel (Lawless and Ramsden 1990). Indeed, the rapid technological innovation, diversification into its downstream uses and agglomeration economies which led to expansion throughout much of the nineteenth century resulted in Marshall (1919) using Sheffield to exemplify his notion of an 'industrial district', a geographically well-defined area with a level of cooperation and flexibility uncommon in other industrially active areas.

This economic growth based upon manufacturing industry continued until the inter-war years, when over-capacity resulted in rationalization in steel and heavy engineering. This temporary problem, however, was soon overcome as Britain rearmed for the Second World War (Taylor et al. 1996). The post-war years, however, were another matter. The major steel producers witnessed severe contraction from the 1970s onwards. This resulted from increasing steel production capacity abroad, the heavy regulation of EU steel production, a decrease in domestic demand and the poor performance and competitiveness of British steel firms (Aylen 1984, Lawless 1986, Morgan 1982, Lawless and Ramsden 1990). Steel industry jobs in Sheffield fell from some 45000 in 1971 to around 4700 in 1993 (Taylor et al. 1996).[1] The Lower Don Valley, once the principal source of jobs and economic activity in the city, and one of Britain's most important and prosperous industrial areas until the mid-1970s, was badly hit (Dabinett 1991, Sheffield City Council 1987, Taylor et al. 1996).

In 1971, there were 287000 jobs, of which 48 per cent were in manufacturing industries in the Sheffield Travel-to-Work-Area (TTWA). By 1984, there were just 235000 jobs, only 27 per cent of which were in manufacturing (Dabinett 1991). Not only was the service-job growth, as in many other provincial cities, insufficient to compensate for the manufacturing jobs lost (Hall 1987, Lawless and Ramsden 1990), but manufacturing employment declined faster than nationally and service growth was weaker (Foley and Masser 1988, Lawless and Ramsden 1990). In consequence, unemployment in the city rose from 4 per cent in 1978, which was below the national average of 6 per cent (Dabinett 1991), to 2 percentage points above the national average during the 1990s (Taylor et al. 1996).

By 1991, therefore, just 22.7 per cent of employees worked in manufacturing. Services, meanwhile, employed 71.3 per cent (see Table 11.1). This does not mean, however, that Sheffield can be described as a post-industrial city. The relative growth of the service sector is largely due to the decline of primary, manufacturing and construction industry jobs, as was the case with Glasgow,

Table 11.1 Sectoral distribution of employees in employment in Sheffield, 1981–91

	1981			1991			% change 1981–91
	No.	%	LQ	No.	%	LQ	
Primary manufacturing and construction	103500	42.2	1.07	62300	28.7	0.99	−39.8
Services	141900	57.9	0.94	154700	71.3	1.01	+9.0
Producer services	19200	7.8	0.91	19600	9.0	0.83	+2.1
Consumer services	83900	34.2	1.05	95300	43.9	1.09	+13.6
Mixed producer/ consumer services	24200	9.9	0.77	25700	11.8	0.86	+6.2
Public services	14600	5.9	0.83	14100	6.5	1.03	−3.9

Source: Census of Employment 1991

described in the previous chapter. Nor, moreover, and again as in Glasgow, has the growth of producer services been strong, expanding by just 2.1 per cent between 1981 and 1991 compared to a national growth rate of 28.4 per cent. Instead, Sheffield is perhaps better characterized as a city increasingly dominated by consumer services. This sector witnessed both absolute and relative job growth in Sheffield during the 1980s whilst the so-called engines of growth upon which consumer services are supposedly reliant suffered severe contraction, thus casting grave doubts on the conventional perception that this sector is dependent upon other industries. Indeed, consumer services grew faster in absolute terms than any other sector of the economy so that by 1991, 43.9 per cent of all Sheffield employees worked in this sector. Today, therefore, Sheffield's 'key' industries are not to be found amongst large manufacturing employers. As Table 11.2 shows, in 1993/4, the four largest employers were all consumer or public services. This trend, moreover, is becoming accentuated. By 1995, according to Rouse (1995), the largest manufacturing companies had further reduced their workforces: Twil by 26.3 per cent compared with 1993/4, Carclo Engineering by 17 per cent, Sheffield Forgemasters by 15 per cent and Avesta by 17 per cent. Some of the largest consumer service industries, meanwhile, had expanded their workforces, such as the University of Sheffield by nearly 10 per cent (University of Sheffield 1995). It is not solely as large employers, however, that consumer service industries dominate the local

Table 11.2 Largest employers in Sheffield, by size of workforce 1993/4

Organization	Sector	No. of employees
Sheffield City Council	Local government	22500
Royal Hallamshire Hospital	Medical services	5380
Sheffield University	Higher education	4272
Northern General Hospital	Medical services	4000
British Telecom	Telecommunications	3500
Twil	Wire manufacturer	3103
Sheffield Hallam University	Higher education	2934
Mainline	Public transport	2806
Carclo Engineering	Engineering, steel	2752
Royal Mail	Postal services	2500
Department for Employment	Government agency	2300
Midland Bank	Banking	2150
Sheffield Forgemasters	Engineering	2026
Avesta	Engineering	1759

Source: Derived from Leigh (1995)

economy. In 1994/5, the £159.6 million turnover of the University of Sheffield (University of Sheffield 1995) was exceeded, according to Rouse (1995), by just four manufacturing companies: SIG (£291m), TWIL (£275.8m), Henry Boot and Sons (£184m) and Carclo Engineering (£174m). Consumer services, therefore, are major and important players in this city. With job decline in the traditional manufacturing base, very weak growth in producer services and expansion in consumer service industries, Sheffield thus appears to be undergoing a transition from a manufacturing city to a place dominated by consumer services.

Neither does this trend seem likely to reverse in the near future. As Table 11.3 illustrates, many of the fastest-growing service sectors in Sheffield are those consumer service industries most frequently defined as dependent activities in conventional discourse (for example medical services, retailing and sport). In a bid to move towards an explanation for this apparently independent growth in consumer services, the next section examines the nature and extent of growth in this sector of the economy.

CONSUMER SERVICES IN SHEFFIELD

Attempts to recast 'steel city' into a new mould began in 1981 when an Employment Department was formed in the City Council. Its

Table 11.3 Fastest-growing service sectors in Sheffield, 1981–91

		% increase in employment
1	Personal services (other)	156.2
2	Owning/dealing in real estate	130.0
3	Medical practices	129.0
4	Other retail distribution (non-food)	121.4
5	Business services	119.9
6	Tourist offices/other community services	116.4
7	Dental practices	94.4
8	Hotel trade	91.1
9	Sport/other recreational services	76.2
10	Cleaning services	75.9
11	Restaurants, snack bars and cafés	60.8

Source: Census of Employment 1991

approach was essentially interventionist, campaigning against closure and rationalization, promoting public-sector employment, and initiating and supporting projects for socially-useful production (Bennington 1986). From 1986 onwards, however, and as in many other cities, a partnership model emerged (Strange 1993). As Lawless (1994) highlights, the major vehicle for achieving this was the Sheffield Economic Regeneration Committee (SERC).[2] Its approach has been based upon a targeted set of sectoral strategies. What distinguishes Sheffield's approach from those of other localities is that it has focused upon consumer services (for example sport, cultural industries, retailing) in its economic development strategy (Sheffield City Council 1995a). This, however, was not a positive choice. Confronted by rapid and severe deindustrialization coupled with very weak growth in the producer services sector, the city was forced by economic circumstance to focus upon this sector at the supposed bottom of the sectoral hierarchy. Sheffield had few other avenues open to it. Indeed, two events in particular lured the city's policy-makers down this road: the application for planning permission in 1986 to construct the Meadowhall regional shopping centre; and the invitation in 1986 to host the 1991 World Student Games. As Dabinett (1991) argues, both resulted in a move away from creating a coherent strategy for manufacturing and towards a consumer services-led approach to economic revitalization. Here, therefore, each of the fastest-growing consumer services and the principal industries used to facilitate local economic development in Sheffield are examined: tourism; sports; cultural industries; retailing; higher education; and medical services. The intention, in so doing, is to explore whether consumer services can indeed take a lead role in local economic development.

Tourism in Sheffield

Compared with other cities, Sheffield was fairly late in adopting tourism as a tool for economic regeneration, with the first tourism strategy being adopted only in 1990. As Bramwell (1993) notes, the principal catalyst underlying the decision to emphasize tourism as a vehicle for economic regeneration was the city's selection in 1987 to host the 1991 World Student Games (discussed in detail below). To promote tourism, Destination Sheffield Limited was founded in 1991. This partnership agency has two objectives: to improve the city's image so as to attract inward investment, tourism and other economic activities; and to attract and service business and leisure tourists (Bramwell 1995). To do this, Sheffield has adopted an event-led promotion strategy which targets specific sports and cultural events and to a lesser extent major anniversaries, celebrations of achievements and conferences and exhibitions (Bramwell 1995).

Such an approach, however, has had its problems. On the one hand, its two objectives of city marketing and city tourism marketing are frequently mutually exclusive, as has already been shown in the context of Glasgow in the previous chapter. As Bramwell (1993) asserts, city marketing and city tourism marketing often involve different target markets and marketing channels and sometimes also different products and place images. Improving the city's overall image, for example, is targeted towards inward investors and the government, whilst city tourism marketing targets specific visitor groups.

On the other hand, Sheffield is short of hotels, especially at the top end of the market. In 1991, for example, the first new hotel in Sheffield for 25 years was built. Despite this, Sheffield still only has 2000 beds in city hotels and guest houses. The result is that hotel occupancy rates have steadily risen from 56.1 per cent in 1993 to 62.4 per cent in 1994 (Bramwell 1995), second only to York in the Yorkshire and Humberside region (Heeley 1996). Although a new 128-bed hotel is opening in 1997 and another 210-bed hotel is planned, the problem is that an event-led strategy leads to intensive demand at certain peak periods when a major event takes place. The Euro '96 soccer championships, for example, brought some 30000 football supporters, journalists and officials to Sheffield, causing acute accommodation shortages, dealt with by creating mass camp-sites.

Despite such difficulties, Sheffield does appear to have had some success with its event-led approach. In 1992, and excluding Meadowhall shoppers, the city attracted 3.3 million visitors, spending an estimated £145 million in total (National Heritage Select Committee 1995). By 1993, this had risen to 4 million visitors, spending £170 million (Destination Sheffield Visitor and Conference Bureau 1994). The result, as shown in Table 11.3, is that tourist-related enterprises such as hotels, tourist offices, restaurants, snack bars and cafés are among the fastest-growing service enterprises in the Sheffield economy. To understand how and why this is occurring, the two principal consumer services industries which are driving this event-led strategy are now examined: the sports sector and the cultural industries.

The sports sector in Sheffield

Sheffield is a city steeped in sporting history. Sheffield FC is the world's oldest football club and Hallam FC play on the oldest ground. To explore the extent to which the sports sector is enabling the revitalization of Sheffield, however, the story begins in the mid-1980s. With the formation of SERC, the public and private sector realized that they had a common desire to develop Sheffield but had different ways of achieving it. So, they looked for something that they could do together to improve its economic position and image. At this time in 1986, the Director of Leisure in the City Council received a letter from the British Students' Sports Fédération (BSSF), the British section of the Federation Internationale du Sport Universitaire (FISU), inviting bids from UK cities to host the World Student Games. The opportunity was seized (Roche 1992, 1994).

Sheffield was chosen by the BSSF in February 1987 to go forward as the British nomination to FISU and was awarded the Games in November 1987. Whilst no other bids were received by FISU, it was widely suggested within Sheffield that the city's preemptive bid staved off competition from other nations.[3] The

Universiade, as it is formally known, is a festival of sport and culture held every two years since 1923. This was the first time that Britain had hosted it. The Games were to be the launch pad for the 'new' Sheffield. In 1986, when the Council was searching for a flagship project that would dispel the traditional equation of Sheffield with muck, grime and metal-bashing, the idea of hosting the Games seemed ideal. The total cost of the facilities was £147.1 million, the principal components being Ponds Forge International Sports Centre (£50.8 million), Sheffield Arena (£33.7 million), the Don Valley Stadium (£28.1 million), Hillsborough Leisure Centre (£12 million) and the £11.7 million refurbishment of the Lyceum Theatre (Short *et al.* 1990).

What, however, were the economic impacts of the Games and the facilities developed for them? As with most sports impact studies, the analyses were prospective rather than retrospective (Foley 1991, Short *et al.* 1990). As Table 11.4 reveals, these estimated that the Games would create 6583 (direct and indirect) jobs in the locality at a cost of approximately £24780 per job (£82410 if direct jobs alone are considered). This average cost per job compares favourably with other projects. Foley (1991) estimates that the average cost per job in 1990 of jobs created by Development Corporations, Industrial Improvement Areas, the Urban Programme and Regional Policy was £28760. Over the ten years after the event, moreover, it was calculated that there would be 4068 new permanent jobs in the Sheffield area (6645 in Yorkshire and Humberside) as a result of the facilities, mostly in the tourist industry (Short *et al.* 1990).

Great emphasis, furthermore, was placed on the intangible impacts (for example the enhanced image of Sheffield and the

Table 11.4 Economic impact of World Student Games

Activity	Local Impact[a] No. of jobs (cost/job)	Regional impact[b] No. of jobs (cost/job)
Construction	5504 (£24760)	9739 (£14000)
Organisation and visitor impact	1079 (£24840)	1333 (£20105)
Total jobs	6583	11072
Average cost/job	£24780	£14730

Source: Foley (1991)

Note: a The locality is defined as within 10 miles of Sheffield
   b The region is defined as the standard region of Yorkshire and Humberside

resulting potential inward investment). Indeed, these intangible impacts were one of the principal reasons for deciding to host the Games (Price 1988). Unfortunately, when the Council committed itself to the World Student Games, nobody realized that they were unsaleable to television. The result was a £157 million bill for building the facilities to be paid over the next twenty years (Hamilton-Fazey 1993a), limited commercial sponsorship and an overrunning of their budget by £5 million (Hamilton-Fazey 1993b). Moreover, all this was well publicized nationally, which reinforced, rather than altered, Sheffield's negative image as a place in which not to invest (Bramwell 1993).

Nevertheless, it is perhaps still too early to assess the impacts of the Games. This is because the intangible (as well as the direct and indirect) economic impacts are often not realized until long after the event (see chapters 5 and 6). Although Sheffield appears to be just as unsuccessful today as it was at any time prior to the Games in attracting inward investment, with just 418 jobs created in 20 companies between 1991 and 1996 (Yorkshire and Humberside Development Agency 1996), so far as using the facilities to host further events is concerned, there has been much greater success. For instance, the 1990 McVitie Challenge athletics meeting at the Don Valley Stadium attracted 22000 spectators, the largest attendance at an athletics meeting since 1963 (Markie 1990). Since the Games, moreover, more than 250 national and international sporting events have taken place attended by 700000 people (of which ten have been world championships and six European championships) and 76 have received national or international television coverage, adding up to an estimated £85 million of free publicity (Waple 1996). Such one-off events, however, are not the only means by which the sports facilities have been used.

As in the USA, the facilities have also attracted regular sporting events by professional teams who have made their home in Sheffield. The Arena, which in 1995 attracted 850000 visitors (Destination Sheffield 1996), is the home of a championship-winning ice hockey team (Sheffield Steelers), whilst Ponds Forge, attracting 916000 visitors in 1995, was the home of the championship-winning men's and women's basketball teams (Sheffield Sharks and Hatters respectively) until they moved to the larger Arena. The Don Valley stadium, moreover, is the base for the Sheffield Eagles rugby team. Although no formal studies have been conducted of their economic impacts, there is a perception that they have done much to restore civic pride. The ice hockey and basketball teams regularly attract attendances of 9000 and 6000 respectively in spectator sports previously unpopular in Britain, but sports in which Sheffield is winning. In a bid to build upon this success, initiatives are currently underway to build a £6 million National Basketball Centre (next to the Arena) and a £60 million Sports Institute to provide sport-related research and education. Opening in 1997, moreover, is a £20 million leisure complex next to the Arena to capture more spending from visitors to the Arena and the Don Valley stadium. Indeed, the National Heritage Select Committee (1995), in an inquiry into the future of major sporting events in Britain, concluded that despite the stadiums and sports centres built for the World Student Games costing £20 million a year in debt charges, the Council's decision to use the Games as a means of stimulating urban regeneration appears to have succeeded.

Whether Sheffield will follow the example of Indianapolis in successfully using sport as a tool for local economic development, however, remains to be seen (see chapter 6). The preliminary indicators, however, signify some success. In 1991, 2.7 per cent of employees in employment in Sheffield worked directly in the sports and other recreational services sector (SIC 979), compared with just 1.55 per cent in Britain as a whole. Sheffield, therefore, has a larger than average market share of the sports sector. Neither are these mostly part-time jobs for women, as is sometimes assumed. Some 71 per cent are full-time and 78 per cent held by men. Indeed, between 1981 and 1991, the number of part-time jobs in this sector remained static, whilst full-time jobs increased by 150 per cent. Nevertheless, it may still be too soon to make any judgements. Key questions remain. What could the money have been spent on if it had not been spent on the Games? Is this value for money? In Sheffield, many believe that there were few other options at the time (Price 1988). Although this might be true so far as manufacturing and producer services are concerned, it was obviously not so with regard to other consumer service industries, as the next section reveals.

Sheffield's cultural industries

The 1980s saw a growing recognition that the cultural industries could contribute to local economic development on account of their ability to generate external income for localities, prevent local income from seeping out of the area and change the image of a place. To achieve this, as chapter 9 illustrated, various approaches have been adopted. Cultural industry *strategies* have been either relatively producer- or consumer-oriented, whilst their *goals* range from integrationist to promotional. Although most localities pursue an approach somewhere near the middle of both spectra, Sheffield is widely heralded as one of the principal cities which has pursued a production-oriented strategy with integrationist goals. However, although this might have been the case in the early stages of its cultural industries strategy, this section will show a shift towards a consumption-oriented approach with promotional objectives in recent years as it has been recognized that both consumption- and production-oriented cultural industries are required to create the synergies necessary for successful local economic development in this sector.

In Sheffield, and arising out of research into the music sector of the economy, a cultural industries strategy was first adopted in the mid-1980s, when the decision was taken to develop a Cultural Industries Quarter (CIQ) located within an historic cutlery area of the city centre. The intention was to build upon the established cultural industries already present, including the Leadmill (a live music, theatre and dance centre), which opened in 1982, and the Yorkshire Artspace Society, which housed 26 businesses in jewellery, sculpture, photography, painting and pottery. The first development was Red Tape Studios, which opened in late 1986. This municipal recording and rehearsal studio offers both training and professional facilities to musicians, especially the young unemployed, to develop their musical skills and was in keeping with its production-oriented approach and integrationist goals. So too was Audio Visual Enterprise Centre (AVEC), a £2.5 million development, housing three state-of-the-art 24-track commercial recording studios, Sheffield Independent Film, Tele-Video productions (suppliers of news and sports coverage to a number of broadcasters) and the Untitled photographic gallery. The focus, therefore, was firmly upon commercially viable production-oriented cultural industries, such as recording studios, independent record labels and tele-video productions (Sheffield City Council 1995b). Reflecting these developments, film production, distribution and exhibition showed a 25 per cent increase in employment in Sheffield between 1981 and 1991. Meanwhile, employment in libraries, museums and art galleries decreased by 36.9 per cent, showing the increase in production-oriented cultural industry jobs but decline in consumption-oriented industries (Census of Employment 1991). By 1991, nevertheless, only 0.46 per cent of employees in Sheffield were employed in cultural-related industries (SIC 971-7), compared with 0.82 per cent in Britain as a whole. Employment in this sector, therefore, still remained well below the national average. Not least, this is perhaps because the neighbouring city of Leeds was already a dominant force and regional centre for precisely those media-related cultural industries which Sheffield was attempting to develop (see chapter 12).

This production-oriented approach, nevertheless, continued into the early 1990s. In 1992, the Workstation, a £3.5 million managed workspace complex, opened next door to AVEC. Tenants include the Northern Media School, the International Documentary Film Festival, the Community Radio Association, the regional offices of the Independent Television Commission, the Forced Entertainment theatre group, Yorkshire and Humberside Screen Commission, Sheffield Independent Film, Total Sound Solutions and the Women's Cultural Club as well as graphic designers Eg.G, and Atkin Wilde (Sheffield City Council 1995b).

Post-1992, however, there have been a number of more consumption-oriented developments. In 1995, a £3 million Showroom art house cinema opened, followed in 1996 by a media library, education suite, exhibition and conference facilities and a wealth of restaurants and cafés in the CIQ. Alongside this, a number of night-club developments took place. In a bid to attract clubbers from outside the city, a highly successful 'Dirty Stop Outs' campaign was launched in 1995 which arranges transport to enable them to tour the various clubs. Further reinforcing these consumption-oriented developments, 1997 witnessed the opening of a £12 million Live Arts Centre, with bars, shops, artspace, a jazz café, theatre and studio. The major flagship project in this new consumer-oriented strategy, however, is the proposed £10 million National Centre for Popular Music due to open in 1997, an exhibition centre which traces the history of popular music. The only similar museum is the Rock and Roll Hall of Fame opened in 1995 in Cleveland, Ohio. What is apparent, therefore, is a shift from a production-focused to a consumer-oriented strategy in the CIQ.

There has also been a gradual switch in the goals of such development. Although the early CIQ industries had specifically integrationist objectives (for example Red Tape Studios), the projects are now more promotional, such as the National Centre for Popular Music. The unsuccessful attempt to attract the Royal Armouries Museum to Sheffield from the Tower of London, which switched to Leeds after failing to agree terms with Sheffield, was similarly typical of this flagship-based consumer-oriented approach. So too is the current idea of developing a Millennium building to house surplus exhibits from the Victoria and Albert Museum, expected to attract 250000 visitors annually.

This shift away from production-oriented strategies based upon integrationist goals and towards promotional consumer-oriented strategies through flagship projects can be explained to be the result of two processes. First, it is due to the recognition that CIQs have to combine both production- and consumption-oriented cultural industries in order to succeed. CIQs solely based on production industries, for example, may fail to create the lively atmosphere required to attract cultural producers, consumers and further investment. Second, it is the outcome of a wider shift in the goals of cultural development. The early approach, based on integrationism, can be seen as a legacy of the 1970s, when cultural industry strategies had a political slant, especially in New Left authorities like Sheffield, who sought to develop the alternative cultural sector (for example Red Tape Studios). The recent trend towards a more promotional approach, however, can be understood as a shift towards an approach which puts greater emphasis on the role of cultural industries in external income generation and economic development. Whatever the reasons for this trend, however, Sheffield's cultural industries strategy, like that of many other cities, has diversified into downstream uses, as did its manufacturing industry a century or so earlier. The outcome is that by 1995, there were 770 employees in 150 firms with a £12 million annual turnover in its CIQ. This, moreover, is expected to rise to 3500 jobs in 400 companies by the year 2000 (Sheffield City Council 1995b). These consumption-focused, promotional and flagship-oriented trends in its cultural industries, moreover, are mirrored in Sheffield's retail sector.

Retailing in Sheffield: Meadowhall Regional Shopping Centre

Meadowhall, a £230 million development, is one of just five RSCs in Britain (see chapter 8). After some competition between Rotherham and Sheffield to attract this complex, Meadowhall was given outline planning permission in 1987 and opened in 1990 next to Junction 34 of the M1 on the site of the old Hadfield steelworks in the Lower Don Valley. Nine million people live within one hour's drive. It is composed of 284 units, including six 'anchor' stores (Marks and Spencer, C & A, Debenhams, House of Fraser, Boots and Sainsbury's) and the Oasis, one of the largest open-plan food halls in Europe, with 1000 seats.

Since opening, it has proved an immensely popular attraction both to Sheffielders and outsiders. On average, over 480000 people use it each week (Taylor *et al.* 1996). It is not the case, however, as is sometimes claimed, that its users are the affluent middle classes who have abandoned city centres. Howard and Davies (1992) find that although relatively affluent social groups are slightly more likely to use Meadowhall than poorer social groups, the clientele is not so biased towards the affluent as some commentators would have us believe (for example Rowley 1993). Indeed, at Meadowhall, what is remarkable is its all-inclusiveness, rather than the minor differences between particular social groups in their tendency to visit this complex (Howard and Davies 1992). Neither is there poor access for those without private transport: 19 per cent of Meadowhall visitors use public transport (Howard and Davies 1992). Indeed, consumer research reveals that on the issue of 'ease of access', Meadowhall is rated as highly as other centres within a 30-minute drive time of this complex and customer profiles show that lack of access to a car makes no difference to whether Meadowhall is visited (Howard and Davies 1992). The problem is perhaps that critics confuse access to RSCs with access to free-standing superstores and retail parks, which is poor (Bromley and Thomas 1993).

What, however, have been the economic impacts of Meadowhall on Sheffield? Based on a sales turnover of £325 million per annum (Fieldhouse 1996) and the fact that 30 per cent of trade is drawn from those living over 30 minutes' drive time away (Howard and Davies 1992), at least £97.5 million is generated in external income. Given that there are 7000 jobs at Meadowhall, approximately 2100 of these will be extra jobs which would not have existed in Sheffield if Meadowhall had not been built. These, moreover, are solely the jobs in the RSC itself. Excluded are the many additional jobs indirectly created in associated activities and those induced from successive rounds of expenditure out of the incomes produced from the direct and indirect jobs. Its external income-generating function is not the only economic benefit of Meadowhall. This complex also prevents income from leaking out of Sheffield both by reducing the need for the local population to go elsewhere to shop (for example Chesterfield, Manchester, Leeds) and by encouraging consumers to shop in locally owned businesses, since RSCs have a higher proportion of local tenants than the traditional high street (Williams 1992a, 1995).

More intangibly, it can be argued that Meadowhall also impacts on the image of Sheffield to both potential investors and the local population. So far as the former is concerned, Meadowhall has had both a magnetic effect, functioning in much the same way as the key industry in a conventional manufacturing growth pole by encouraging other business to relocate in its environs (for example the Abbey National and a major retail warehouse development), and a demonstration effect, displaying that businesses can successfully locate in the Lower Don Valley. It has also promoted civic boosterism. With the rise in shopping as a leisure pursuit, such icons of the good life make an area seem an attractive place in which to live, work and do business not only to potential investors but also to the local population, whose sagging pride was in desperate need of an uplift following wave after wave of manufacturing disinvestment in the city (Williams 1992a, 1995). As highlighted in earlier chapters, however, such intangible impacts are easy to assert but seldom measured or valued. Future studies will thus need to pay greater attention to assessing these intangible impacts.

Despite these direct, indirect and intangible economic impacts, Meadowhall is still perceived by some as a pariah which has, amongst other things, decimated the city centre (Rowley 1993). Even before Meadowhall, however, the city centre was in a bad state of disrepair and in decline. If anything, the result of the competition from Meadowhall has been the rejuvenation of Sheffield city centre after years of benign neglect. In order to promote competition, £500 million was spent on major improvements to the city centre in the early 1990s.[4] Neither has there been an outflow of stores to Meadowhall. All of Meadowhall's anchor stores are present in the city centre and in many cases have undertaken extensive refurbishment of their premises, spending £15 million between 1993 and 1995, and those living in Sheffield know that there are many factors besides Meadowhall which have inhibited trade in the city centre during the 1990s, not least the construction of Supertram. In other words, the self-inflicted 'push' factors seem just as important to any exodus of people as the 'pull' factor of Meadowhall itself (Williams 1995).

Meadowhall, therefore, far from being a 'parasite', has been a key factor in the revitalization of Sheffield. It has brought many new jobs and a better-quality shopping environment in Sheffield, and has helped to reconstruct positively the city's image. Indeed, it is precisely due to its success that other neighbouring cities are developing their own RSC (for example the White Rose Centre in Leeds and the Trafford Centre in Manchester) in order to be able to compete effectively with Sheffield.

Sheffield Universities

The University of Sheffield and Sheffield Hallam University are the third and seventh largest employers in the city respectively. The University of Sheffield, for example, employs 4694, has an income of £159.6 million and 13916 full-time and 1903 part-time students (University of Sheffield 1995). Indeed, of the 230000 people employed in the Sheffield TTWA, 2.8 per cent are employed in higher education, compared with 1 per cent nationally (Census of Employment 1991). Sheffield, therefore, is more heavily reliant on higher education than most localities and has the sixteenth highest concentration of higher education employment of all Local Labour Market Areas in Britain (Goddard et al. 1994).

What, therefore, are the economic benefits of this over-representation of higher education jobs in Sheffield? Although no formal economic impact assessments have been conducted on either university, it can be asserted that they generate jobs both by bringing in external income and through the multiplier effect created by local purchasing of goods and services by the universities themselves, their employees and their students (see chapter 7). With a combined income of £223 million in 1993/4, 7206 employees and 29230 full-time equivalent students (Leigh 1995), the universities make a substantial contribution to the health of the Sheffield economy. It is not solely in terms of the external income-generating function and resulting multiplier effect, however, that they contribute to local economic development. Universities also have intangible impacts through the support they provide to local companies in the form of technology and non-technology transfer, new spin-off firm formation and their impacts on inward investment.

To enable technology and non-technology transfer, the University of Sheffield has set up a Business Development Service to help forge strategic business collaborations and partnerships as well as a Regional Office which, for example, has founded Project Link University of Sheffield (PLUS) to give students the opportunity to conduct projects in local organizations and in 1994/5, handled projects involving 875 students. In total, the University of Sheffield received industrial support in 1994/5 of £2.8 million (University of Sheffield 1995). Hallam, meanwhile, has a large number of research centres and short courses providing both technology and non-technology transfer, such as the Advanced Manufacturing Technology Advice Centre to advise local small and medium-sized companies on various aspects of manufacturing. In addition, Sheffield has 12 Teaching Company Schemes and Hallam 16. Both universities, moreover, have policies to encourage spin-off companies and Hallam has a university-sponsored science park, with 30 companies employing 200 workers. In 1993/4, Sheffield claimed four spin-off companies employing seven staff whilst Hallam simply declare one or more companies (Leigh 1995).

Less tangibly, there are also the contributions to local economic development which result from the university staff's involvement in local governance (for example on Sheffield Development Corporation, Sheffield TEC, school governorships). Goddard *et al.* (1994) find that Hallam have 20 staff members who are District/County councillors, 5 on TEC Boards, 2 on Health Authorities/Trusts, 12 school governors and 20 company directors. Besides such economic impacts, these universities also have a range of social and community impacts. Both Sheffield and Hallam offer children's activities, drama productions, public lectures and sports facilities for the public, but only Sheffield offers local history and music whilst Hallam alone offers art exhibitions (Leigh 1995). Take, for example, the contribution to the cultural vitality of the area. The Sheffield Cultural Workshop, which involves both universities and the City Council, brings together performers, policy-makers and university staff with the aim of encouraging cultural activities in the region and serving as a forum for the discussion of new ideas. Hallam also has a large media school which creates synergies with the CIQ.

In sum, there is little doubt that the two universities in Sheffield not only are major contributors in themselves to the economic vitality of Sheffield but also make a range of indirect and intangible contributions to Sheffield. It is important to note, however, that these university/locality relationships are not a one-way dependency relationship. Rather, there is inter-dependency. Take, for example, technology and non-technology transfer. Although these universities facilitate such transfers, they are also dependent on the local economy for their continuing success and income. Without investment from local concerns, many of the transfer activities and the accompanying research would cease. Similarly, the success of these universities in attracting students is dependent not solely on the reputation of the universities themselves but also on the image of the city. Sheffield's recent ascendency as a 'clubbing capital' in the North because of its cultural industries strategy, for example, will have direct impacts on the attractiveness of the universities to potential students, displaying how the vitality and viability of different sectors in this local economy are inter-dependent. This synergy is nowhere more prominent than in the medical services sector.

Medical services in Sheffield

The Royal Hallamshire Hospital, employing 5380, and Northern General Hospital, employing around 4000, are the second and fourth largest employers in Sheffield. Indeed, between 1981 and 1991, employment in hospitals, nursing homes, etc. (SIC 951) rose by 51.6 per cent so that by 1991, 6.8 per cent of all employees in Sheffield worked in this sector compared with 5.4 per cent nationally. Medical services (for example medical and dental practices), in particular, were amongst the fastest-growing sectors in the city. Until now, there has been little analysis of the contribution of either hospitals or medical services to local economic development. However, those studies which have been conducted, albeit in a North American context, suggest that this sub-sector plays a similar role in economic development to the other consumer services considered throughout this book. Hospital services, that is, are a major external income generator and also facilitate income retention (Gross 1995, Hiles 1992). Such services, therefore, are far from being dependent activities.

In a bid to facilitate greater income generation and prevent leakage of income by creating stronger inter-linkages between the various sectors which constitute the medical industry in Sheffield, the Medilink initiative was launched in 1995. This brings together local private-sector medical manufacturing and service companies, hospitals, universities and business support agencies to facilitate innovation in this sector through collaboration. To do this, four strategic business support services have been established: research and product development; information exchange; marketing and export; and training. The research and product development group, for instance, are working on translating new ideas which arise from the region's clinical base into products through collaboration between local companies and agencies. In the orthopaedics sector, for example, such collaborations are being used to initiate new developments and market Sheffield as a centre for orthopaedics (Liversidge 1995). Consequently, medical services, like many other consumer services, appear to be functioning as a motor for local economic development and through their inter-relations with other sectors of the local economy, are promoting economic regeneration.

CONCLUSIONS

Sheffield, in sum, has undergone a rapid transition. Once the heart of global steel manufacturing, it is now a city dominated by consumer services. However, the conventional perception of consumer services as dependent activities is far from the reality in contemporary Sheffield. Whilst the so-called engines of growth (i.e. primary and manufacturing activities) have severely contracted, consumer services have rapidly grown. Indeed, tourism, sport, the cultural industries, retailing, higher education and medical services are all making positive contributions to the regeneration of Sheffield not only in their much-neglected role as 'basic activities' but also in their non-basic locally-oriented role of preventing the seepage of money out of the area.

It is little surprise, of course, that in steel city, there have been voices decrying the pursuit of consumer service jobs at what is seen as the expense of 'real' jobs. Dabinett (1991: 13), for example, argues that there has been 'an opportunistic emphasis on leisure, sport and retailing associated with *ad hoc* attraction of inward investment . . . Policies for major local industries such as metal processing and engineering have, on the other hand, been under-developed and unfocussed.'[5] However, given the widespread decline of the traditional industries, a radical rethinking of Sheffield's economic base was necessary. The signs indicate that to some extent, Sheffield is making a success of its consumer service-oriented sectoral approach.

Indeed, just as in an earlier era, Sheffield became a powerhouse of the national economy by developing an industrial district around its core manufacturing activities, based upon technological innovation, cooperation, flexibility, diversification and agglomeration economies, today, there appears to be sufficient evidence to suggest tentatively that many of these trends are now being replicated in its consumer services sector. On the basis of technological innovation (for example in the development and marketing of medical services, advanced retail facilities, tourism products, cultural industry services and sports-related activities and products), a high level of cooperation (for example Medilink, the CIQ initiative, sports-related developments) and agglomeration economies (for example in the CIQ and Meadowhall), Sheffield can be asserted to be replicating the conditions in its consumer services sector which made its manufacturing base so successful in a previous era.

However, this consumer services-led strategy is not without problems. As the retailing, sporting and cultural developments have arisen mostly from opportunism rather than strategic planning, the resulting product mix is both structurally diverse and geographically diffused. There is, for example, no single strong theme or major visitor attraction which is of such importance that it alone can entice large numbers of visitors to the city. Neither is there a distinctive district, with facilities being geographically dispersed and not within walking distance of each other (Bramwell 1993). Nor is there even a clear focus. Although it is apparent to many that cultural industries, sport and tourism, for example, are all central thrusts of the economic development strategy, they remain relatively fragmented policy arenas distinctly lacking an overall strategy to provide inter-linkages and collaboration between them.

Related to this, the major focus remains an outward-oriented approach which seeks to use these sectors to induce income into the economy and serve the wants of people outside the area. In comparison, there is an under-emphasis on leakage prevention and developing products and services which meet local demand. Similarly, there is little evidence that this consumer services focus has done much to reduce the socio-economic and spatial polarization created by the previous rounds of manufacturing investment and disinvestment (Lawless 1995a and b). However, and despite such problems, there is little doubt that Sheffield exemplifies how consumer services can play a lead role in local economic revitalization.

12

CORPORATE CITY

The role of consumer services in the rejuvenation of Leeds

INTRODUCTION

From the soot-blackened West Riding city dominated by heavy engineering and textiles, usually associated with gritty kitchen-sink dramas, Leeds has come through a fundamental sea-change where the 'rags to riches' epithet is, for once, accurate.

(Green 1994: 75)

Leeds, the industrial heart of West Yorkshire, shines out as a role model for other northern cities – a powerhouse of regeneration blessed with the lowest unemployment rate of any English city, despite being the only metropolitan area to increase its population between the 1981 and 1991 censuses.

(Wavell 1995: 37)

After London and Birmingham, Leeds is the third largest city in England, with a population of some 717000 in 1991. Although for a long time, this major urban area failed to project either a positive or a negative image of itself onto the national psyche, in recent years, as the above quotations intimate, a more positive picture has started to emerge. Leeds has become widely acknowledged as a successful urban economy, encapsulated in headlines such as 'Leeds leads the way' (Rice 1995) and 'an economic hothouse with amazing resilience' (Green 1994). This upturn in the fortunes of the Leeds economy is popularly explained as being due to the growth of producer services in general, and financial services in particular (for example Rice 1995, Tickell 1996). The aim of this chapter, however, is to evaluate critically whether this is indeed the case. As Hall (1995) asserts, a considerable number of sub-global and regional cities have recently witnessed a rapid

185

economic resurgence based on their ability to attract investment in both producer and consumer services. Is Leeds unique, therefore, in the sense that its revival is based on producer services alone? Or is it simply the case that the contributions of consumer services to the rejuvenation of Leeds have been so far disregarded? To investigate this, first, the chapter will profile the structure of the Leeds economy so as to analyse the growth of producer and consumer services, and second, it will evaluate the role of consumer services in the revitalization of this regional city through an exploration of the two consumer service industries which have been taken on board in local economic policy: the cultural industries and retail distribution.[1] This will reveal that for all the emphasis on producer services in explanations for the growth of Leeds, consumer services have also played a significant and crucial role in its regeneration. Indeed, and in contrast to Glasgow and Sheffield, discussed in the previous chapters, although local economic policy-makers have cast such services in a primarily supportive role in local economic development, this chapter will reveal that they have played additional roles. To commence, however, and so as to explain the different way in which consumer services are conceptualized in Leeds compared with Glasgow and Sheffield, the direction and character of the Leeds economy are examined.

THE LEEDS ECONOMY

At the time of its incorporation as a city in 1893, Leeds had already developed into one of the major industrial centres in Britain. Coal mining, engineering and tanning were dominant industries, followed by clothing and printing. Innovations in these sectors coupled with specialization had enabled Leeds to expand rapidly (Thomas and Shutt 1996). As the twentieth century progressed, nevertheless, the innovative capacity of locally owned firms deteriorated as they became incorporated into the subsidiaries of externally owned companies. The result was rapid deindustrialization. Indeed, by 1991, manufacturing employed just 20 per cent of employees in the city, the vast majority in externally owned and controlled firms (Thomas and Shutt 1996). This deindustrialization of Leeds, however, is not solely due to dwindling innovative capacity. With relatively high land and labour costs in this city compared with neighbouring areas, and regional assistance unavail-

Table 12.1 Comparative employment change in major British cities

	% change in employment 1981–91	% unemployed February 1995	% change in no. unemployed 1983–94
Leeds	4.0	7.7	−22.9
Cardiff	−8.5	10.0	−27.0
Birmingham	4.3	8.9	−16.2
Glasgow	−10.9	9.6	−39.6
Bristol	3.6	7.9	−2.5
Manchester	−10.5	8.7	−21.4
Newcastle	1.0	10.7	−23.9
Sheffield	−12.4	10.5	−22.4
Liverpool	−23.3	12.8	−33.2
Edinburgh	4.1	6.9	−20.6
Nottingham	−0.1	9.8	−3.5
Leicester	−5.3	7.4	−17.4

Source: Derived from Campbell (1996: Table 3.4)

able to firms in Leeds since the 1960s, the restructuring of Leeds towards the service sector is far from a simple outcome of global restructuring forces. Rather, it reflects the interplay between global trends, local conditions and national policy regimes (Haughton and Williams 1996).

In comparison with other major urban areas such as Glasgow and Sheffield, nevertheless, deindustrialization has not resulted in severe economic problems. Between 1981 and 1991, whilst most other major cities lost large numbers of jobs, the size of the workforce in Leeds increased by 4.0 per cent, and the number of unemployed fell by nearly a quarter between 1983 and 1994. Indeed, by 1995, the unemployment rate was lower than in any other major city in Britain, except Edinburgh (see Table 12.1). Such a performance is made all the more remarkable when it is recognized that Leeds was the only metropolitan area to increase its population between 1981 and 1991 (Campbell 1996, Wavell 1995).

Why, therefore, has Leeds witnessed such outstanding employment growth relative to other major British cities? The answer, in short, lies in the service sector. Whereas in most British cities, service-job growth has been insufficient to compensate for the fall in manufacturing employment, this has not been the case in Leeds. As Table 12.2 reveals, between 1981 and 1991, the loss of 26100 primary and secondary jobs was more than counteracted by the gain of 39900 service-sector jobs. The outcome has been that the city has quickly shifted from an over-reliance on manufacturing compared with the national economy to an over-representation of service jobs.

This transformation of Leeds into a service-based economy, nevertheless, is not merely due to the growth of producer services. Although the over-representation of producer services reinforces the perception that the resurgence of Leeds is due to this sector, and financial service employment grew by 68 per cent between 1981 and 1991, making Leeds the fastest-growing regional financial centre (Tickell 1996), other services have also witnessed significant growth. From 1981 to 1991, consumer service employment expanded by 26.3 per cent so that by 1991, 39.4 per cent of all employees worked in this sector. Indeed, for every producer service job created between 1981 and 1991, 2.7 jobs were created in consumer services (a 25700 net increase in consumer service jobs compared with 9500 in producer services). Nonetheless, despite such growth, consumer services still remain under-represented in Leeds when compared with the national economy. This is surprising given its role as a regional capital. The reason will be discussed below. What should not be inferred from this under-representation, however, is either that consumer services are insignificant in the recent resurgence of Leeds or that their importance is diminishing over time.

Indeed, examining the major sources of job growth in Leeds, Table 12.3 shows that many consumer service industries are amongst the most rapidly expanding sectors in this local economy. Although business services created 11100 net jobs and banking and finance 5100 jobs between 1981 and 1991, the combined effects of the employment growth in the consumer service sectors of medical services, education, social welfare and community services and retail distribution far outstrip the job gains in these producer services. Consumer services, therefore, have played a major part in the expansion of this burgeoning service-oriented economy. Furthermore, and as Table 12.4 illustrates, the largest employers in this city are no longer the chief manufacturing corporations and neither, for the most part, are they producer service firms. Instead, the vast majority of the principal employers are consumer service industries.

Consequently, the growth of Leeds is not simply due to the

Table 12.2 Sectoral distribution of employees in employment in Leeds, 1981–91

	1981			1991		
	No. of jobs	*% of jobs*	*Location quotient*	*No. of jobs*	*% of jobs*	*Location quotient*
Primary, manufacturing and construction	114300	38.1	0.97	88200	28.1	0.97
Services	185600	61.9	1.01	225500	71.9	1.01
Consumer services	97800	32.6	0.99	123500	39.4	0.98
Producer services	29800	9.9	1.15	39300	12.5	1.16
Mixed producer/consumer services	35900	12.0	0.87	48300	15.4	1.12
Public services	22100	7.4	1.04	14400	4.6	0.73
Total Employees	299900	100.0		313700	100.0	

Source: Census of Employment 1991

Table 12.3 Largest job gains in Leeds: by sector, 1981–91

Sector	Net job gains
Business services	11100
Medical services	7600
Education	5800
Social welfare/community services	5200
Banking and finance	5100
Retail distribution	5000
Sanitary services	2800
Wholesale distribution	1700
Insurance	1100
Postal services and telecommunications	800

Source: Census of Employment 1991

Table 12.4 Eleven largest employers in Leeds, 1993/4

Organization	Sector	No. of employees
Leeds City Council	Local government	37000
Leeds General Infirmary	Medical services	6000
St James's Hospital	Medical services	5200
University of Leeds	Higher education	4733
Royal Mail	Postal services	3900
Leeds Metropolitan University	Higher education	2509
Dept of Social Security	Public services	2300
British Gas	Energy supply	2200
Yorkshire Electricity	Energy supply	1800
First Direct	Banking	1600
Leeds Permanent Building Society	Financial services	1500

Source: Extended from Leigh (1995)

expansion of producer services. The consumer services sector is a principal component of the net increase in jobs in this city, and many of the fastest-growing industries as well as the largest employers are consumer services. The question which remains unanswered, however, is whether the upturn in the fortunes of Leeds is due to both consumer and producer services as Hall (1995) suggests, or whether it is producer services which are playing the lead role and consumer services are simply expanding by living off the endeavours of this sector.

CONSUMER SERVICES IN LEEDS

Examining the discourse surrounding local economic policy in Leeds, one gets a clear sense that producer services are held to be the principal motor of local economic development. Since the mid-1980s, and as in most other cities, the strong streak of city father paternalism and municipal socialism which characterized governance in Leeds has been swept aside on the tide of public/private partnership (Haughton 1996). The principal vehicle through which this local corporatism has been embraced is the Leeds Initiative, launched in 1990. A sectoral approach was adopted, and the first project, instigated in 1993, was the Leeds Financial Services Initiative (Tickell 1996), reflecting the embeddedness of the view that producer services are the main engine of local economic growth. Although later initiatives by this self-named 'corporate city' have targeted retailing (as well as engineering, printing, clothing and textiles, health care and medical instruments, reflecting the locality's manufacturing roots) and, at the time of writing, further sectoral initiatives are planned in the cultural industries (as well as construction and wood products), the overwhelming impression is that the approach towards consumer services in local economic policy in Leeds is very different from that in Glasgow and Sheffield, discussed in the previous two chapters.

In Glasgow and Sheffield, as shown, consumer services have been placed at the heart of economic policy and given a basic-sector role, mostly because few other options were open to them. In Leeds, by contrast, where there has been a good deal of success in attracting producer services, consumer services have been cast in a different role. If the literature produced by the economic development agencies is examined, it becomes immediately apparent that although they are not relegated to a 'dependent' or 'parasitic' status, neither are they given prominence as a lead sector. Instead, they are primarily viewed in a supportive role as bolstering the city's attractiveness to other sectors. Here, however, an examination is undertaken of whether this supportive role is their only function in local economic development, or whether they are also playing additional roles. To do this, the two specific sub-sectors of consumer services which have been targeted by the Leeds Initiative for development will be analysed: the cultural industries and retail distribution.

191

The cultural industries in Leeds

As chapter 9 demonstrated, the cultural industries are increasingly becoming an integral part of many economic development strategies. Although Leeds has not focused upon the cultural industries with the same fervour as other cities (for example Glasgow and Sheffield), mostly because it weathered deindustrialization better, the cultural industries have still emerged as an important constituent of the city's approach towards local economic development. Indeed, establishing Leeds as a major social and cultural centre is a key objective of both the City Council's economic strategy (Leeds City Council 1992) and the Leeds Initiative (see Haughton 1996). What approach, however, has Leeds adopted towards the cultural industries? Does it see the cultural sector as simply bolstering the attractiveness of Leeds to external investors? Or does it view the cultural industries as engines of growth in their own right? To answer this, the particular approach adopted towards the cultural industries in Leeds will be examined here using the framework developed in chapter 9, which characterizes cultural industry strategies as either production- or consumption-oriented and as pursuing either locally oriented integrationist or externally oriented promotional goals. To understand the approach adopted in Leeds, the types of cultural industry being cultivated will be explored, and an analysis of the goals behind such development will follow. This will then enable conclusions to be drawn about whether this sector is seen simply as a vehicle for promoting the attractiveness of Leeds to other sectors or whether it is also cast in a lead-sector role.

At the outset, it is important to state that there is no explicit cultural industries strategy in Leeds at the time of writing. This strategy is still in its planning stages. Nevertheless, and despite this lack of a formal strategy, a distinctive approach can be discerned. To see this, one needs only to examine the range of cultural industry developments in the city. Leeds is the home of Yorkshire Television, BBC North-East, two radio stations, more than twenty independent film, television or video producers, and ten recording studios (Leeds City Council 1992). As such, it is the regional centre for media-related cultural industries. Despite this, and perhaps because of the pre-existing strength of these production-oriented cultural industries, it is the consumption-oriented cultural industries which have been the focus for cultural industry policy and development. Building upon a strong performing-arts base in the city, including Opera North, the West Yorkshire Playhouse, the Phoenix Dance Company, and the Northern School of Contemporary Dance, the Northern Ballet Theatre relocated in 1995 from Halifax to Leeds along with its sister school, the Central School of Ballet. Leeds also hosts an International Piano Competition and has an annual concert season including performances by brass bands, string quartets and chamber group. In addition, there are a number of popular arts and cultural events which are regular features in Leeds' cultural calendar, such as Opera in the Park and Party in the Park (an open-air popular music concert), Jazz on the Waterfront, Rhythms in the City (a five-week summer festival of music and performance on the streets of Leeds), the St Valentine's Fair in the city centre and the Leeds West Indian Carnival in Chapeltown, second in size only to the Notting Hill Carnival in London.

The museums sector has also been a principal focus of development efforts, such as the conversion of a former Victorian workhouse adjacent to St James's Hospital into Europe's largest medical museum using £3 million of lottery funding. The major flagship project in this realm, however, has been the relocation of the surplus artifacts from the Royal Armouries in London to the £42.5 million Royal Armouries Museum, which chose Leeds after failing to agree terms with Sheffield, and is the first purpose-built national museum to be constructed outside London.

Leeds has also embarked on high-profile campaigns to transform its image by becoming both a '24-hour city' and a 'European' city (Smales and Whitney 1996). These encompass the liberalizing of the city's licensing regulations and fostering a café culture, which results in the city's streets being once more populated out of regular office hours (Leeds City Council 1994). In particular, the intention here is to encourage after-work socializing given the number of young professional workers commuting into Leeds each day. Related to this, the city can now boast a number of the country's leading nightclubs and a vibrant youth culture (Wainwright 1995). In 1991, there were 2507 jobs in nightclubs and licensed clubs in Leeds, 456 more jobs than the national average for an area of this size (Williams 1996b). Of course, some of these additional jobs will be due to the fact that Leeds, as a major city, attracts people into its environs from surrounding smaller towns and cities, such as Wakefield, Barnsley and Bradford. However, a further proportion will have been created by Leeds competing effectively with cities such as Manchester and Sheffield for 'clubbers', an increasingly competitive market in which customers have a finely attuned perception of whether a city is meeting their desires. Its development, moreover, has many 'knock-on' effects for the growth of other economic sectors. A principal reason for many students deciding to come to Leeds to study has been its 'clubbing' reputation, which reveals the way in which the various sub-sectors of consumer services are closely inter-connected.

One problem in Leeds pursuing the '24-hour' and 'European' city notions, however, is the lack of restaurants, snack bars and cafés. There are 3837 jobs in such businesses, 351 jobs fewer than the national average according to the minimum requirements technique (Williams 1996b). Again, one might consider that since it is a regional capital, which acts as a magnet for night-life and a focus for business, this sector would be over-represented in Leeds. It is not. How, therefore, can this be explained? One reason is that in comparison to neighbouring cities such as Bradford, a 'cultural identity' for its restaurant sector has not been developed in a bid to draw people into Leeds to spend their money. Thus, whilst Bradford has managed to entice people from Leeds to visit its restaurants, with its marketing appeal as 'curry capital of the North', there has been little coordinated and managed response from Leeds itself. If progress is to be made towards the 'European 24-hour city', then a more coordinated approach to this sub-sector of consumer services will be required.

These initiatives reveal, nevertheless, that Leeds adopts a predominantly consumption-oriented approach to cultural industry development rather than a production-oriented perspective. Indeed, this consumption-oriented focus is being further reinforced by the current attempt to create a cultural industries quarter. The City Council's submission to the Millennium Fund for a £70 million arts and entertainments quarter at Quarry Hill proposes to include the expansion of the West Yorkshire Playhouse, the Northern Ballet Theatre and the Central School of Ballet, the Leeds College of Music, the Film Theatre of the North, the Yorkshire Dance Centre and the Phoenix Dance Company, a media workshop, the National Blues Museum and a small concert hall for up to 500 people. These are all primarily consumption-oriented facilities. Little attempt has been made to develop production-oriented cultural industries, in stark contrast to Sheffield, despite Leeds being a regional centre for sectors such as the media-related industries. This is reflected in the cultural industries employment which has resulted. Between 1981 and 1991, authors, music composers, etc. (SIC 976) increased by 496.2 per cent and librarys, museum and art gallery jobs by 30.6 per cent, but radio/television services rose by just 7.2 per cent (perhaps because this sector was already firmly established in Leeds by 1981), whilst film production, distribution and exhibition decreased by 20 per cent, reflecting this consumption-oriented focus (Census of Employment 1981 and 1991).[2]

It is fairly clear, therefore, that Leeds adopts a consumption-oriented approach towards the cultural industries. Whether this intimates that cultural industries are supported more as vehicles for improving the attractiveness of Leeds to other sectors than as a lead sector itself can only be answered by examining if the principal goal is an externally oriented promotional one and if so, whether it is primarily attracting cultural tourists or seeking investment in other sectors by making Leeds attractive to them.

Superficially, a promotional approach does not appear to have been pursued. Instead, an integrationist one seems to have been adopted. Not only are both popular as well as high arts promoted, but the philosophy appears to be that expense should not prohibit anybody from attending an event. Indeed, events such as Opera in the Park and Party in the Park are free, as are the majority of the concerts in the City Council's concert season. There are also concessionary fares at most other venues. However, although such events appear superficially to be integrationist in orientation, they are also promotional because the large number attending guarantees media attention and displays the city as having a vibrant cultural life. It is thus not pure altruism on the part of the authorities. Indeed, the way in which cultural industry development is approached by the Council and the projects pursued relate much more to promotional goals, and this seems to have dominated the cultural industry agenda increasingly over time. As Strange (1996a: 148) concludes from a survey of those involved in cultural industry development in Leeds, 'For some involved in arts-based community work in Leeds, the arts as voice for the marginalised has diminished as its economic and image-building potential has been exploited.' Epitomizing this emphasis on promotional goals is the Royal Armouries Museum. Opened in 1996, this flagship project symbolizes Leeds's attempt to become a major arts centre,

a European city and a centre of tourism. It is expected to transform the city's tourist industry, creating between 850 and 1000
permanent jobs, attracting a million visitors per year and generating
£10–15 million for the city economy (Green 1994). Adjacent to the
museum is Tetley's Brewery Wharf, a £6 million attraction telling
the story of the British pub developed by Joshua Tetley and Son
attracting 250000 visitors annually.

The result, therefore, is that as a more coordinated approach
develops towards the cultural industries, with the Leisure Services
Department of the Council and the Leeds Initiative acting as the
catalysts, a sharper focus appears to be emerging which emphasizes
externally oriented promotional goals. As Strange (1996a: 150)
asserts, 'the current approach to cultural policy in Leeds is largely,
although by no means exclusively, promotional'. This consumption-oriented promotional approach is pursued, moreover, primarily as a way of enhancing the image of the city to make it more
attractive to inward investment. As one local councillor put it,

> Top business people will tell you how important the arts are in
> their decision making. The DSS made the decision to relocate
> to Leeds because of the arts and cultural life of Leeds. . . . The
> thing that distinguished Leeds from other places was the
> cultural life of Leeds. They couldn't believe that the cultural
> life was so rich. We've found that with a number of major
> organisations which have moved here.
>
> (Strange 1996a: 140)

Recognizing this role of the cultural industries in local economic
development, the Leeds Initiative has at the time of writing identified the cultural industries as a future sectoral initiative for attention in local economic policy. It appears, therefore, that
promotional goals are increasingly prominent, and these primarily
involve attracting investment in other sectors rather than simply
enticing cultural tourists. The conclusion, therefore, is that the
cultural industries are mainly seen to play a supportive role in local
economic development as a vehicle for making Leeds attractive to
investment in other sectors rather than as a lead sector itself.

Despite this, there is little doubt that the contributions of the
cultural industries to the economic revitalization of Leeds are both
as a leading and as a supporting sector. The cultural industries,
employing 14000 people in 500 companies (Strange 1996b), are not
only creating the conditions to facilitate inward investment in other

sectors but also generating external income. This consumer service industry, far from being parasitical, is therefore an 'engine of growth' in the Leeds economy both as a lead and as a supportive sector. Can the same be declared, however, of the retail sector?

Retail distribution in Leeds

As highlighted in chapter 8, retailing more than any other consumer service industry is endowed with a 'parasitic' role. In Leeds, nonetheless, this is not the case. It is viewed as a principal vehicle through which the city can promote itself as an attractive place in which to live, work and do business. Here, however, it will be shown that although this is the only role attributed to it in policy, as has been argued before, this sector does in fact have a larger role to play. Here, the issue is examined of whether the retail sector is also a basic-sector industry, generating external income for the city, and a potential means of preventing the leakage of money out of the area.

Over the past decade or so, Leeds has witnessed considerable investment in its retail core in a bid to stave off competition from neighbouring areas and make Leeds attractive as a location for investment in other economic sectors. 'Landmark Leeds' is the name given to the £3.6 million redevelopment of the pedestrian retail zones in the city centre, which were first created in the 1960s. This infrastructure renewal encompasses decorative paving, landscaping, new street furniture, lighting and a series of gateways to act as focal points. Three modern indoor shopping centres (St John's Centre, Schofields Centre and Bond Street Centre) have also been constructed along with the refurbishment of the Victorian arcades and the Merrion Centre, the first indoor shopping centre to be built in Britain. In addition, the converted Corn Exchange opened in 1990 as a speciality retail centre and in 1996, a £2 million scheme was completed to restore the 1875 Victorian market, which stands alongside the recently revamped Edwardian market hall at Kirkgate Market. In 1996, moreover, Harvey Nichols opened their first store outside London in Leeds, providing another large anchor store to add to those already present, including two House of Fraser stores – Rackhams and Schofields – Marks & Spencer, Debenhams and Lewis's.

The consequence of such retail investment is that between 1981 and 1991, retail jobs increased by 13.2 per cent in Leeds TTWA compared with a national increase of just 12.2 per cent. Despite this, by 1991, at 10.1 per cent, the proportion of employees in Leeds's retail sector was still below the national average figure of 10.7 per cent. Leeds, therefore, was under-represented in retail employment, even though it is the major shopping centre for the region and has increased its market share.[3] Nevertheless, there is little doubt that in terms of the inter-urban competition for retail spending, Leeds has performed well. As Williams (1994c) has documented, the increasing concentration of West Yorkshire retail spending in Leeds is draining ever greater amounts of money out of the lower-order neighbouring centres of Huddersfield, Bradford, Wakefield and Halifax. The result is that these latter centres slipped down the national league table in terms of their position in the retail hierarchy, judged by the number of multiples present in each locality, from 27th, 130th, 86th and 61st in 1971 to 43rd, 139th, 90th and 78th in 1989 respectively. Leeds, meanwhile, rose from 8th to 6th over the same period (Hillier Parker 1991) and now has 4.3 million square feet of retail floorspace, 950 retail outlets and at 625000 square feet, the second highest demand in Britain for retail floorspace. Indeed, between 1991 and 1995, the net output in retailing increased from £491 million to £550 million, a 12 per cent increase (Leeds TEC 1995).

Consequently, the retail sector in Leeds appears to be performing a basic-sector function. It is generating greater amounts of external income by persuading outsiders to spend their money in the city. This, in turn, creates multiplier effects which produce yet further jobs in the local economy beyond the retail distribution sector itself. In an era of competitive advantage where cities are rivals across all industrial sectors, no sector can escape from having to contend with its equivalent in other towns and cities. This applies as much to retailing as any other sector. In Leeds, this competition comes from two sources: a host of out-of-town regional shopping centres; and established city and town centres such as York, Manchester, Sheffield, Wakefield, Bradford and Harrogate. Each will be considered in turn so as to assess the effectiveness of the Leeds retail sector as an income generator.

A number of out-of-town regional shopping centres are within driving distance of Leeds. Some 30 miles to the south of Leeds is Meadowhall in Sheffield, 90 miles to the north is the MetroCentre and 40 miles to the west is the Trafford Centre in Manchester. One significant response taken by Leeds to this type of competition has been to build its own regional shopping centre. The 650000-square-foot White Rose Centre at Churwell, which opened in 1997, has Debenhams, Bhs and C & A as anchors. If Leeds had not captured this complex, it is likely that one of the other neighbouring towns or cities would have done so, with resultant impacts on retail employment in Leeds. Nevertheless, there is little concern amongst retailers in Leeds city centre about its deleterious effects (Rice 1995). In major part, this is due to the strength of retail investment in the city centre in recent years and the relatively small size of this RSC: its retail floorspace is just 15 per cent of the total currently available in the city centre.

So far as other competition from established city and town centres is concerned, although retail employment in Leeds is growing faster than in its West Yorkshire neighbours, there is increasing competition from North Yorkshire shopping centres such as York and Harrogate. This is due to the way in which affluent households are spatially distributed within the Leeds conurbation. Not only are the vast majority of affluent areas within the city located in the northern suburbs, but an increasing proportion of the relatively affluent who work in Leeds live outside the metropolitan district boundary in towns and villages to the north of the city (Stillwell and Leigh 1996). A consequence is that Leeds is increasingly in competition with the relatively up-market shopping centres of York and Harrogate for these affluent shoppers. The problem, therefore, and this partially explains the low levels of retail employment in Leeds (and perhaps consumer services more generally), is that a proportion of people working in this city both live outside and shop elsewhere, which results in a high leakage of income out of the Leeds economy. An indirect response was for the Leeds Initiative to set up the Leeds Retail Initiative in 1994 so as to promote city centre shopping. Recognizing that a more coordinated approach was required to make Leeds as a shopping centre competitive (Smales and Whitney 1996), it has adopted a pro-active approach to retaining its position as a first-order centre in the North and staving off the competition.

In sum, the retail sector has not only been successful as a key vehicle for transforming the image of Leeds so as to make it attractive to potential investors but has also gained market share over its neighbours in West Yorkshire and thus generated external income. In this sense, the Leeds retail sector is far from being the dependent or parasitic sector often portrayed in conventional economic development discourse. Neither, however, is it solely an industry which bolsters the attractiveness of Leeds for investment in other sectors. Instead, it also has an income-generating role, acting as an engine of growth for the Leeds economy. One cautionary note, nevertheless, is that with the location of affluent socio-economic groups in the north of the city and their out-migration, and the increasing competition from York and Harrogate for their custom, Leeds is suffering to some extent from a leakage of income out of its economy. Given that it is net income, rather than total external income, which results in economic growth for a locality, this is a problem which will need to be tackled in the future. At present, and on account of the perception that retailing has only a supportive role, such a problem is not being seriously addressed in local economic policy.

CONCLUSIONS

To return to the question posed at the start of this chapter, the revitalization of the Leeds economy, similar to many other sub-global and regional cities, is due not only to the growth of its producer services but also to the rapid expansion of its consumer services. So, despite all the emphasis on producer services when explaining the remarkable success of Leeds, this chapter has revealed that the consumer services sector is the source of most of the net increase in jobs in this city and that many of the fastest-growing industries and largest employers are consumer services. This sector, therefore, is playing a significant and crucial role in its regeneration.

Evaluating the nature of this contribution by examining the two specific industries which have been the focus of local economic policy towards consumer services (the cultural industries and retail distribution), this chapter has revealed that such consumer services not only create the conditions necessary for attracting investment in other sectors but also act as a lead sector because of their income-generating role. Up to now, however, although local economic policy has recognized that consumer services are not parasitic activities and has realized that they are important in their supportive role of attracting investment, their role as income generators has been relatively ignored. To a large extent, this is because unlike other cities such as Glasgow and Sheffield, Leeds has been relatively successful in attracting other basic industries, such as producer services, which means that it has not had to resort to consumer services to the same extent as these other cities.

However, this lack of recognition of the full contribution of consumer services to local economic development has led to at least one major problem in the Leeds economy. Because of the failure to focus upon the role of consumer services in improving net income, little attention has been paid to this sector as a vehicle for preventing the leakage of money out of the economy. The result is that Leeds has a poor performance in preventing such seepage. Many people who work in Leeds live outside the city's metropolitan district boundary, and this has led to a loss of income to the city. By 1991, 68000 per day commuted into Leeds to work (22 per cent of the total labour force), compared with 51800 in 1981. The number commuting out, meanwhile, has risen much more slowly, resulting in 31000 more people commuting to Leeds than outwards in 1991, compared with a figure of just 17300 in 1981 (Leeds Development Agency 1994). The consequence is that there is a considerable and growing leakage of money out of the Leeds area. Until now, this has been relatively ignored. Greater emphasis on curtailing this seepage, nevertheless, would produce a major boost for the Leeds economy. To achieve this, much greater consideration needs to be given to the leakage-preventing abilities of the consumer services sector. The re-creation of the night-time economy in Leeds is one step in this direction. The success of a corporate city, after all, is as dependent upon retaining money within its walls as it is upon generating additional external income. Leeds, nevertheless, has a long way to go in mending some of these cracks so as to prevent the trickle outwards turning into a flood.

13

CONSUMER SERVICES AND RURAL REGENERATION
A case study of the Fens

INTRODUCTION

Rural economies in advanced nations have undergone a radical transition in the post-Second World War period, shifting away from a heavy dependence on agriculture and becoming more diversified. In the US and the UK, for example, just 8 per cent and 6 per cent of all rural jobs respectively are in agriculture, whilst manufacturing employs well under a quarter of rural workers (US Department of Agriculture 1993, Department of Environment and Ministry of Agriculture, Fisheries and Food 1995). The result is that the vast majority of rural employment is now in the service sector. Indeed, this has been the case for more than two decades (Glasmeier and Howland 1994a and b). Despite this, most rural research and economic policy remain focused on the primary and secondary sectors. As for their urban counterparts, therefore, there has been an under-emphasis on the role of services in rural economic development.

Even when the service sector is addressed, moreover, the stress has been on its weakness compared with urban areas. The share of overall employment in services has been shown to be lower than in urban areas (Bluestone and Hession 1986, Glasmeier and Howland 1994a and b, Tarling et al. 1990), and many producer service industries have been revealed to be strongly urban in orientation, for example advertising, banking, insurance and computer services across the advanced economies (Coffey and Polese 1987, Marshall et al. 1987). When this is coupled with the fact that deindustrialization, although widely prevalent across the urban landscape, is absent in rural areas, [1] the result is that major reservations have been voiced about the role of services in rural economic development (Glasmeier and Borchard 1989, Glasmeier and Howland 1994a and b).

Here, however, the intention is to investigate whether services in general, and consumer services in particular, can be written out of the rural economic policy quite so easily. To do this, first, the way in which consumer services are portrayed as 'dependent' activities in rural discourse will be evaluated critically and second, the contributions of consumer services to rural economic development will be explored through an empirical study of the rural economy of the Fens, a hinterland area in East Anglia. Investigating the magnitude of external income generated by manufacturing industry, producer services, consumer services and mixed producer/consumer services in this rural economy, along with the extent to which these sectors use local suppliers, this will reveal that consumer services are playing an active and beneficial role in the development of this rural economy not only in their much-neglected role as basic activities which attract external income but also in their non-basic locally oriented role of curbing the seepage of income out of the area. The outcome, it is argued, is a need for the rural studies literature to modify its perspective on the role of such services in rural economic development.

CONSUMER SERVICES AND RURAL ECONOMIC DEVELOPMENT

Rural economies, like urban areas, have traditionally pursued economic development by attempting to increase the amount of external income flowing into their localities. In so doing, they have conformed to the orthodox interpretation of 'economic base' theory, that an area needs to attract external income in order to grow. Rural economic development, therefore, like urban economic policy, has focused upon developing its basic industries. Indeed, rural economies can be judged to have been relatively successful in generating external income. As Keeble *et al.* (1992) reveal, rural businesses tend to be more outward-oriented than their urban counterparts: they sell a higher proportion of their total output outside the locality. This is not due to the locality being too tightly defined. Firms in both remote and accessible rural areas sell a significantly higher proportion of their total output outside their region than firms in urban areas.

Which sectors, however, generate external income for rural

economies? In rural studies, the way in which this question has been answered has shifted over time. In the eighteenth century, for instance, a group of French economists called the physiocrats, led by François Quesnay (1694–1774), held that agriculture was the only source of wealth. They asserted that there were three classes of society: landowners; farmers and farm-labourers; and others, called the 'sterile class'. Only the agricultural sector was seen to produce any surplus value, the rest only reproducing what it consumed. As industrialization progressed, however, this essentially agrarian view came under increasing attack. The basic sector was increasingly conceptualized as being composed of both primary and secondary activities, so that service activities alone became classified as 'sterile', reliant upon the primary and manufacturing sectors, which attract external income (Clout 1993, Redwood 1988). This reconceptualization of the basic sector is mirrored in rural revitalization strategies. As Clout (1993) highlights, there has been a steady shift away from a focus upon modernizing agricultural production and towards the inclusion of manufacturing, with grants and loans being made available to encourage manufacturing firms to decentralize their operations to rural areas and to facilitate expansion or modernization by manufacturing companies already in rural areas. Indeed, today, and despite the tertiarization of rural economies, much rural economic policy remains firmly rooted in this conceptualization of the primary and secondary sectors being the sole external income generators.

However, and replicating the shifts in thinking on urban economies, some have sought to show the external income-generating function of both rural producer services (Porterfield and Pulver 1991, Smith 1984, Smith and Pulver 1981, Stabler 1987, Stabler and Howe 1988, 1993) and tourism (Keane and Quinn 1990, Nilsson 1993, Tarling et al. 1990). Scant attention, however, has been paid to other potential basic consumer service industries in rural areas, such as the cultural industries sector.[2] Nevertheless, even if the range of consumer services being considered as basic activities is not wide, at least some elements of the consumer services sector are slowly being recognized in the rural economic development literature as having a role to play in generating external income and thus jobs.[3]

The same, however, cannot be said of their role as leakage preventers. Few rural development agencies consider this issue. A result is that there is a deficiency of effective policy instruments to stem leakage of income out of rural localities. Current local sourcing campaigns, for example, simply seek to put local businesses in touch with each other. These have been largely unsuccessful because it is not solely a matter of imperfect information on the availability of local goods and services which prevents local sourcing. Businesses work within the constraints of the logic of the market where availability, quality and price are the major determinants of the suppliers used. Little, however, has been done by rural development agencies to intervene in the market in this regard (Williams 1996e). The main means of encouraging rural consumers to buy locally, meanwhile, the 'use it or lose it' approach applied mainly to retail facilities, has also been ineffective on account of several intransigent problems. The diversity of the goods and services available in rural areas is typically poorer and the price higher than in urban centres, so people still tend to go to urban areas to shop. When these problems of diversity and price are coupled with lifestyle changes, such as greater car ownership, women's participation in employment and heightened consumer demand for more sophisticated products, the result is that rural people prefer to drive to an urban retail centre less frequently to choose from a wider range of goods where prices are lower (DoE/ MAFF 1995). Indeed, even if people do buy locally, the businesses may often be owned externally and/or the goods and services sourced from external suppliers, so a share of the profits and income will quickly leak out of the area. It will not be recirculated within the rural economy. Furthermore, where these 'use it or lose it' campaigns have been used, they have tended to be seen as social rather than economic initiatives, whose function is to help preserve a sense of community in rural areas rather than facilitate economic rejuvenation (Williams 1996e).

If the contribution of consumer services to rural economic development is to be understood fully, therefore, consumer services must be viewed as basic activities and leakage preventers. These two features, moreover, are as important in rural economic development as in urban rejuvenation since both are required to increase net income. The problem at present, however, is that consumer services are not widely considered as a sector worthy of consideration in rural economic development, with the notable exception of tourism. Here, therefore, and as a corrective to current discourse, their contributions to rural economic development as both income generators and leakage preventers are examined through a case study of the Fens. This area has been chosen for empirical investigation since such a rural hinterland is precisely the type of area in which consumer services are least expected to play a significant role in economic development.

CONSUMER SERVICES IN THE FENS ECONOMY[4]

The Fens, here defined as the designated area of Fenland District Council, is a rural area located between Peterborough, Cambridge and Kings Lynn within East Anglia. Its population is clustered in the small market towns of Wisbech, Chatteris, March and Whittlesey. Throughout the 1980s and 1990s, the county and region of which it is a part has been one of the fastest-growing areas in Britain in terms of population and employment. Examining the nature of the economy, Table 13.1 shows that the Fens, reflecting the high-grade agricultural land of this area, is over-represented in primary and secondary jobs (many of the latter being work in branch plants involved in food processing). The service sector, meanwhile, is under-represented in this economy. Between 1981 and 1991, moreover, although there was a slight increase in service

Table 13.1 Sectoral distribution of employees in employment in the Fens, 1981–91

	1981			1991			% change 1981–91	
	No.	%	LQ	No.	%	LQ	Fens	Britain
Primary, manufacturing and construction	9800	49.0	1.26	9900	42.7	1.47	+1.0	−19.6
Services	11200	56.0	0.92	13300	57.3	0.81	+18.7	+18.6
Producer services	2600	13.0	1.51	2800	12.1	1.12	+7.7	+28.4
Consumer services	6200	31.0	0.95	7700	33.2	0.83	+24.2	+26.0
Mixed producer/ consumer services	1500	7.5	0.58	2300	9.9	0.72	+53.3	+8.9
Public services	900	4.5	0.63	500	2.1	0.33	−44.4	−9.6
All industries	20000	100.0		23200	100.0		+16.0	+2.0

Source: Census of Employment 1991

employment, this did not keep pace with national trends, leading to a lower location quotient in 1991 than ten years earlier.

From this, it can be concluded that the Fens seems superficially to be an excellent candidate for championing the view that services contribute little to rural economic development. This is further reinforced by taking a precursory glance at the nature of services in the Fens. The consumer services sector seems to be composed of little more than lower-order services which supply the local population only. Even the tourist industry is poorly developed, with little evidence of new ventures and enterprises. Producer services, meanwhile, appear just as undeveloped. There is no evidence of back-office functions having transferred into this locality and throughout the 1980s, the Economic Development Units of both Cambridgeshire County Council and Fenland District Council were unable to convince producer service firms, refused sites in Peterborough and Cambridge, to contemplate locating in the Fens. The outcome, as Table 13.1 illustrates, is that between 1981 and 1991, job growth in both producer and consumer services was below the national average, leading to lower location quotients than a decade earlier. It appears, therefore, that any further inspection of the economy will merely corroborate the view that services in general, and consumer services in particular, are dependent activities, reliant upon the primary and secondary sectors in this rural economy.

In a bid to confirm or refute this first impression of the contribution of consumer services to rural economic development in the Fens, a postal survey of all 728 business establishments on the register of commercially rated premises, along with all farms in the Fens, was conducted in March 1990. In all, 270 usable responses were obtained (a 37.1 per cent response rate).[5] The survey examined the markets and suppliers of Fenland businesses.

Geography of markets

Table 13.2 reveals the market orientation of the various industrial sectors in the Fens. This shows that services are more local-market-oriented than manufacturing, thus corroborating the finding of previous studies that the rural service sector is less export-oriented than the manufacturing sector (Porterfield and Pulver 1991, Stabler and Howe 1993). These microeconomic data also uncover, however, the dangers of stereotyping all manufacturing as

LOCALITY CASE STUDIES

Table 13.2 Geographical structure of markets of Fenland businesses: by sector and firm size

| | Source of customers (%) | | | |
	Local[a]	Regional	National	Number of businesses
Primary	56.4	61.4	100.0	7
Manufacturing	39.9	58.3	97.8	84
Metals, minerals and chemicals	14.6	37.4	97.0	7
Metal goods, engineering and vehicles	51.8	75.1	98.6	30
Other manufacturing	35.5	50.2	97.3	45
Construction	52.4	87.9	99.1	18
Services[b]	68.3	84.5	98.3	149
Producer services	51.8	70.7	98.0	54
Consumer services	84.1	92.6	98.0	54
Mixed producer/consumer services	73.1	94.2	99.5	41
All industries	57.4	75.8	98.3	256
Small firms (<20 jobs)	69.2	86.2	99.4	177
Medium firms (20–100)	30.5	53.8	95.2	62
Large firms (>100 jobs)	38.5	51.2	98.3	17
Services				
Small	71.6	86.6	99.4	118
Medium	52.1	75.1	93.4	24
Large	72.3	85.0	100.0	7
Producer services				
Small	58.0	75.7	99.4	44
Medium	31.8	42.0	95.7	7
Large	43.6	64.9	99.9	3
Consumer services				
Small	84.4	93.1	99.8	43
Medium	80.1	83.6	84.4	8
Large	91.7	100.0	100.0	3
Mixed consumer/producer services				
Small	64.0	15.1	20.8	31
Medium	57.8	91.7	99.1	9
Large	100.0	100.0	100.0	1

Note: a 'Locality' is defined as within 20 miles of the establishment; 'region' is defined as East Anglia

b The definition used to categorize service-sector activity into producer, consumer and mixed producer/consumer services is that developed by Marshall (1988: Table 2.4)

Source: Author's survey

'basic' and all services as 'dependent' activities. As Table 13.3 reveals, 11.9 per cent of manufacturing establishments do not export any of their output outside the Fens, whilst 15.4 per cent of the services sell none of their output locally. Hence, the perception of services as entirely locally oriented is wholly inaccurate in the Fens. Only 20.8 per cent of service-sector establishments conform to such a configuration, even when 'local' is defined as within a 20-mile radius. Just as importantly, only 23.8 per cent of manufacturing firms sell all their output outside the locality. In consequence, it is erroneous either to promote all manufacturing industries as basic activities in the Fens or to relegate all service-sector industries to the dependent sector. Indeed, 36.9 per cent of

Table 13.3 Proportion of customers in Fenland businesses who are local: by sector and firm size

| | % of businesses with customers of whom | | | | | |
	0% local	>25% local	>50% local	>75% local	100% local	No.
By sector						
All manufacturing	23.8	42.8	36.9	28.6	11.9	84
All services	15.4	75.2	66.4	57.7	20.8	149
Producer services	22.2	59.3	46.3	38.9	16.7	54
Consumer services	9.2	85.2	79.6	74.1	31.5	54
Mixed producer/ consumer services	17.1	78.0	73.2	53.6	14.6	41
Large						
Consumer services	0.0	100.0	100.0	100.0	0.0	3
Producer services	0.0	66.6	33.3	33.3	33.3	3
Mixed producer/ consumer services	0.0	0.0	0.0	0.0	100.0	1
Medium						
Consumer services	16.7	83.3	83.3	83.3	16.7	8
Producer services	42.8	28.6	28.6	28.6	14.3	7
Mixed producer/ consumer services	12.5	75.0	62.5	37.5	0.0	9
Small						
Consumer services	8.9	84.4	75.6	68.9	33.3	43
Producer Services	18.2	93.2	54.5	43.2	15.9	44
Mixed producer/ consumer services	12.2	58.5	53.6	46.3	12.2	31

Source: Author's survey

Fens manufacturing establishments have more than half of their customers in the locality, whilst 33.6 per cent of services have over half of their customers outside the locality. This study also reveals that international trade in service outputs is not purely an urban phenomenon. In the Fens, just 7.1 per cent of manufacturing firms have customers abroad, compared with 9.4 per cent of service-sector establishments. This finding in a British rural area thus supports the findings from North America that services can function as basic industries in rural communities much as they do in metropolitan areas (Gillis 1987, Smith 1984, Stabler 1987, Stabler and Howe 1993, Porterfield and Pulver 1991).

Therefore, just as in a previous era Adam Smith countered the prevailing discourse emanating from the physiocrats when he opposed the view that agriculture was the only truly 'productive' activity by arguing that manufacturing should also be considered 'productive' (Illeris 1989a), it is perhaps the case today that a new concept of the basic sector is required in rural studies. The evidence suggests that it is not acceptable for merely the primary and secondary sectors to be considered as constituting the basic sector of rural economies. Services can also act as basic industries. Just as importantly, and as has been shown, manufacturing establishments are not always basic in orientation, a fact which should not pass unnoticed.

To analyse further the type of service industries contributing to external income generation in the Fens, Tables 13.2 and 13.3 display the extent to which producer services, consumer services and mixed producer/consumer services each function as basic industries. Whereas 42.9 per cent of all firms have the majority of their customers outside the locality, and 33.6 per cent of all service-sector establishments have such an externally oriented customer base, 53.7 per cent of producer services have the majority of their customers external to the area. Indeed, producer services in the Fens are just as export-oriented as the 'metal goods, engineering and vehicles' sub-sector of manufacturing. Some 22.2 per cent of producer services, moreover, sell all their output externally. This examination of producer services in a British rural area thus verifies the conviction that producer services are basic activities, reinforcing in a British rural context not only the findings of previous British studies which have focused upon producer services in urban areas (Daniels 1991, Marshall 1983) but also the research undertaken on such services in rural communities in

North America (Porterfield and Pulver 1988, Smith 1984, Smith and Pulver 1981, Stabler and Howe, 1988, 1993). So too does it support the notion that rural producer services can have an international customer base. In the Fens, 9.3 per cent of producer service firms have international customers, compared with just 7.3 per cent of manufacturing establishments. In addition, it is not the larger producer service establishments as might be expected, but rather the medium-sized producer service establishments which are most likely to have an international clientele.

This evidence from the Fens, moreover, bolsters the assertion made throughout this book that consumer services can be basic activities, attracting people into the locality in order to benefit from their spending. Consumer services in the Fens, although not so external-income-oriented as either producer services or mixed producer/consumer services, are not 'dependent' activities. As can be deduced from Table 13.3, 20.4 per cent of consumer service establishments sell over half their output to customers outside the locality, thus corroborating that consumer services in the Fens function as basic activities. Indeed, only 31.5 per cent of consumer service establishments correspond with the stereotype for this category of firm of having all their customers within the locality; 68.5 per cent, therefore, do not conform to the pure locally oriented role, as depicted in conventional discourse. Indeed, 9.2 per cent of consumer service establishments have none of their customers within the locality. If this is borne out in future studies of other rural localities, and there is no reason to believe that the Fens is significantly different to other accessible rural areas, then far more weight will need to be given to consumer services as basic activities in the rural economic development literature than has so far been the case.

Take, for example, the retail sector in the Fens. If any sub-sector of consumer services is likely to be a dependent sector serving purely a local market, then it is this industry, which the vast majority of analysts would probably choose. However, of the 16 retail outlets responding to the survey, only 18.8 per cent comply with the dependency-sector stereotype of the retail establishment as serving an entirely local market; 31.3 per cent of retail establishments have the majority of their customers outside the locality. Even in this sub-sector of consumer services, therefore, external income-generating activity is to be found, revealing that no sub-sector of consumer services should be excluded from discussion when it is being considered whether consumer services are basic industries (Williams 1997; see also chapter 8).

Neither is it simply the case that consumer services are serving a slightly wider catchment area than that considered as 'local' or 'regional' in the survey design. In the Fens, 7.4 per cent of consumer service establishments have an international aspect to their customer base, which is slightly higher than for manufacturing establishments, of which 7.1 per cent export abroad. Therefore, the external income-generating activity of consumer services cannot be assumed to be solely a product of how 'local' and 'regional' are defined. So, leaving aside the extent to which they sell their wares to customers from outside the locality and region, this survey finds that they induce foreign income into rural localities, further reinforcing the case for their being considered basic-sector activities.

The outward orientation of consumer service industries, nevertheless, is influenced by both firm size and ownership, albeit not in the way that might be expected from previous studies of the rural service sector (Smith and Pulver 1981). Interestingly, the larger the consumer services establishment in terms of employees, the greater the percentage of customers who are local. However, great care must be taken when drawing such conclusions because of the small number of large and medium-sized consumer services in the survey responses. Indigenous consumer services, meanwhile, are as likely to have external customers as those externally owned consumer services with similar-sized workforces. Although such a finding must again be treated with the utmost caution in view of the small number of externally owned establishments responding, it does contrast with Smith's (1984) assertion that external ownership of services often goes hand-in-hand with export orientation. If these findings in the Fens are replicated elsewhere, this has major implications for rural economic development policy, for it suggests that seeking large externally owned businesses is not necessarily a prerequisite for export activity amongst consumer services.

For some, however, this evidence that consumer services generate external income may be insufficient to persuade them that such industry should be taken seriously as an external income generator. They may argue, for instance, that the extent of external income generation by consumer services in the Fens is not very great and that it is all part of the normal hierarchical system feeding in and out of the surrounding centres, which are larger than those of the study area. To an extent, this may be correct. Nevertheless, it is just as correct for the manufacturing and producer service establishments in this locality. What is important is the net effect. If a sector has a net income effect on the locality, then it will be helping the area grow and develop. It is for this reason that consumer services need to be analysed not only in terms of their external income-generating abilities but also in terms of their role in preventing leakage of money out of the area.

Geography of suppliers

Income generation *per se*, to repeat, is an insufficient but necessary ingredient to engender economic growth in a locality. For an increase in net income to occur, not only external income but also a reduction in total external spending is required. The growth of a rural economy is thus dependent on not only attracting external income but also preventing the leakage of money out of the area (Persky *et al.* 1993). Consumer services, in theory, can diminish the leakage of money by both thwarting the need for the local population to spend money outside the area by providing facilities within the locality, and by sourcing locally to a greater extent than other sectors. Although this survey of businesses is unable to explore the degree to which consumer services reduce total external spending by preventing the leakage of personal expenditure (since this would require a survey of household expenditure patterns), it does provide important evidence concerning the leakage of business expenditure out of the Fens economy (see Tables 13.4 and 13.5).

This uncovers that although consumer services have a smaller proportion of their customers outside the Fens than manufacturing or producer service establishments, they are more likely to source their inputs locally. On average, 44.2 per cent of suppliers to consumer service establishments are local, compared with 37.4 per cent of suppliers to producer services and 27.1 per cent of suppliers to manufacturing establishments (see Table 13.4). Indeed, 35.2 per cent of consumer service establishments have more than half their suppliers within the locality, compared with 31.5 per cent of producer services and 17.8 per cent of manufacturing establishments (see Table 13.5). The outcome is that consumer service establishments will have relatively higher local

213

Table 13.4 Geographical structure of suppliers to Fenland businesses: by sector and firm size

	Source of suppliers (%)			
	Local[a]	Regional	National	Number of businesses
Primary	63.6	86.4	98.5	7
Manufacturing	27.1	45.0	90.1	84
Metals, minerals and chemicals	21.0	34.8	82.6	7
Metal goods, engineering and vehicles	35.1	57.8	93.5	30
Other manufacturing	22.9	38.4	89.0	45
Construction	41.4	71.1	98.3	18
Services[b]	45.3	63.3	95.2	149
Producer services	37.4	56.3	91.9	54
Consumer services	44.2	61.1	96.6	54
Mixed producer/consumer services	61.4	75.7	97.7	41
All industries	39.2	58.3	93.8	256
Small firms (<20 jobs)	43.6	62.9	96.3	170
Medium firms (20–100)	32.0	50.2	87.0	62
Large firms (>100 jobs)	21.6	41.2	93.8	17
All services				
Small	46.2	63.7	97.0	110
Medium	44.7	62.1	86.1	24
Large	31.7	60.0	97.7	7
Producer services				
Small	45.7	63.1	99.1	44
Medium	5.2	21.2	79.2	7
Large	10.6	37.0	98.4	3
Consumer services				
Small	45.6	62.3	98.0	43
Medium	46.5	53.8	91.8	8
Large	30.0	66.7	100.0	3
Mixed producer/consumer services				
Small	67.8	77.1	99.8	31
Medium	81.0	92.8	99.1	9
Large	80.0	90.0	100.0	1

Note: a 'Locality' is defined as within 20 miles of the establishment; 'region' is defined as East Anglia
   b The definition used to categorize service-sector activity into producer, consumer and mixed producer/consumer services is that developed by Marshall (1988: Table 2.4)
Source: Author's survey

Table 13.5 Proportion of suppliers to Fenland businesses who are local:
by sector and firm size

| | % of businesses with suppliers of whom | | | | | |
	0% local	>25% local	>50% local	>75% local	100% local	No.
All manufacturing	33.3	30.9	17.8	11.9	3.6	84
All services	34.9	49.0	35.6	28.9	11.4	149
Producer services	37.0	44.4	31.5	20.4	5.5	54
Consumer services	33.3	44.4	35.2	29.6	9.2	54
Mixed producer/ consumer services	31.7	56.1	48.8	41.5	19.5	41
Large-sized services						
Consumer	33.3	33.3	33.3	33.3	0.0	3
Producer	66.6	0.0	0.0	0.0	0.0	3
Mixed producer/ consumer services	0.0	100.0	100.0	100.0	0.0	1
Medium-sized services						
Consumer	50.0	33.3	33.3	33.3	0.0	6
Producer	71.4	0.0	0.0	0.0	0.0	7
Mixed producer/ consumer services	0.0	87.5	87.5	75.0	25.0	9
Small-sized services						
Consumer	31.1	48.9	42.2	26.7	11.1	43
Producer	29.5	43.2	36.3	27.3	6.8	44
Mixed producer/ consumer services	26.8	39.0	34.1	26.8	14.6	31

Source: Author's survey

multiplier effects than both manufacturing establishments and producer services since a greater percentage of their inputs is obtained from local sources. Mixed producer/consumer services, moreover, source locally to an even greater extent than consumer services and thus have even higher multiplier effects. On average, 61.4 per cent of suppliers to mixed/producer consumer service establishments are local, compared with 44.2 per cent of suppliers to consumer services. Indeed, 48.8 per cent of mixed consumer/producer service establishments have more than half their suppliers within the locality, compared with 35.2 per cent of consumer service businesses.

There is also a strong relationship between local sourcing and both firm size and ownership. As Table 13.5 displays, the larger the business, the weaker is the local sourcing. This applies not only across the service sector as a whole but also to the consumer services sector in particular. Larger consumer service-sector establishments source to a lesser extent locally than smaller consumer service firms. Similarly, when the ownership of establishments is examined, indigenous consumer services are found to be far more likely to source their inputs from within the locality than are externally owned consumer services. The implication for rural economic policy is that supporting smaller indigenous consumer services will result in higher local multiplier effects than if larger externally owned consumer services are enticed into the locality. Indeed, attracting such consumer services into the locality will be detrimental to the rural economy if they replace indigenous small consumer services, for this will decrease the extent of local sourcing.

Moreover, the evidence from the Fens suggests that those sectors that supply the greatest proportion of their output to external customers also source to the greatest extent externally (see Table 13.6). Some 63.1 per cent of manufacturing establishments, for instance, have more than 50 per cent of their customers outside the locality, and 82.2 per cent have the majority of their suppliers external to the economy. Only 20.4 per cent of consumer services, in contrast, have the majority of their customers outside the area and just 64.8 per cent have the majority of their suppliers outside. In between these two sectors are producer service establishments, where 53.7 per cent of such firms have the majority of their customers outside the locality and 68.5 per cent have the majority of their suppliers external to the area.

Table 13.6 Location of customers and suppliers of Fenland businesses

| | % of business with more than 50% of | | |
	Customers outside locality	Suppliers outside locality	Number of businesses
Manufacturing	63.1	82.2	84
Services	33.6	64.4	149
Producer services	53.7	68.5	54
Consumer services	20.4	64.8	54
Mixed producer/ consumer services	26.8	51.2	41

Source: Author's survey

The result is that there is an inverse relationship between the export-orientation of a sector and the level of the local multiplier effects. This rural survey of the Fens thus supports the work of Porterfield and Pulver (1988) in rural areas of the US, which asserts that although service-sector firms export less than manufacturing firms, rural services buy more of their inputs from within the state, offering partial compensation for their poorer external income-generating performance. This has significant implications for rural economic policy. A sector with a strong export-orientation is not necessarily the sector which will most benefit a rural economy. When account is taken of the local multiplier effects, the increase in net income to a locality may be higher amongst those sectors which generate relatively smaller proportions of their total income externally but which source to a greater extent locally. The ramification, therefore, is that the improvement in net income may be greater if a multitude of smaller indigenous consumer services are supported, which are less oriented towards external income generation than other sectors but source to a greater extent locally and retain income which would otherwise be lost, than if relatively more export-oriented manufacturing branch plants are attracted which do not source locally.

Consumer services in the Fens, therefore, are not dependent activities. In this rural hinterland, not only do 20.4 per cent of consumer service establishments have the majority of their customers outside the locality compared with 53.7 per cent of producer services and 63.1 per cent of manufacturing establishments, but also consumer services are likely to have relatively greater local multiplier effects than enterprises in other sectors on account of the higher proportion of local suppliers. Some 35.2 per cent of consumer service establishments have the majority of their suppliers locally compared with 17.8 per cent of manufacturing and 31.5 per cent of producer services. However, great care must be taken when contemplating the wider significance of these findings. Not only does the share of output sold externally by any sector differ across localities, but so too does the extent of local sourcing (Bloomquist 1988, Smith 1984). In consequence, generalizations about the degree to which sectors generate external income or source locally cannot be made from such one-off case studies. What can be stated, nevertheless, is that consumer services are not parasitic activities and should not be ignored in future studies of rural economies.

CONCLUSIONS

This chapter has thus revealed that services are becoming an ever-growing segment of rural economies. Nevertheless, they remain under-emphasized in much rural research and policy-making and when they have been examined, the tendency has been to highlight the weakness of this sector relative to urban areas and to question their importance in rural economic development. Through a case study of the Fens, however, this chapter has displayed the extent to which the service sector in general, and consumer services in particular, contribute to rural economic development in practice. Even in this rural hinterland, which is precisely the type of locality in which consumer services would be least expected to make a contribution to local economic development, consumer services are found to play a positive role not only in their much-neglected role as basic activities which attract external income but also in their non-basic locally oriented role of curbing the seepage of income out of the area.

Rural economic policy, therefore, can no longer dismiss consumer services as dependent activities. Although consumer services do not generate external income to quite the same extent as manufacturing and producer services, this is likely to be more than compensated for by the fact that the inputs into consumer services tend to be locally supplied, thus generating higher local multipliers. Furthermore, it is not solely large externally owned consumer services which generate external income; smaller indigenously owned establishments are equally likely to do so. These smaller indigenous consumer services are also more likely to source inputs locally than large externally owned establishments. Consequently, rural economic policy does not need to attract large external consumer services. Indeed, this may diminish, rather than improve, net income, if they replace indigenous small consumer services. The implication is that an increase in net income is more likely to result in rural areas from developing a multitude of smaller indigenous consumer services, which are less oriented towards external income generation than other sectors but source to a greater extent locally and retain income which would otherwise be lost, than if large externally owned manufacturing branch plants are attracted which do not source locally.

218

14

LONDON

The role and character of consumer services in a global city

INTRODUCTION

One outcome of the new international division of labour based on the spatial separation of corporate functions has been the emergence of global cities. These are the control and command points in the global economy where corporate headquarters, financial intermediaries and highly specialized services agglomerate to convey decisions or to conduct transactions over large areas of the world (Cohen 1981, Friedmann and Woolf 1982, Friedmann 1986, Frost and Spence 1993, Hamnett 1994a and b, Sassen 1991). When global cities are examined, focus until now has been upon producer services in general and financial services in particular (see, for example, Daniels 1993). The aim of this chapter, however, is to explore the character and nature of consumer services in a global city. Are they simply 'dependent' activities or can their role and character be 'read' in a different way, as has been shown to be the case in second-tier cities? Indeed, how does their function differ when compared with the case of the second-tier cities discussed in the preceding chapters?

In a bid to move towards understanding consumer services in global cities, this chapter explores the case of London. As in the preceding chapters, first, the economic restructuring of this city will be examined and following this, the contributions of the consumer services sector to the London economy will be evaluated. This will reveal that far from being dependent-sector activities, consumer services are playing a more important role in global cities than has so far been recognized, albeit different from the one they play in the localities discussed in previous chapters.

219

THE ECONOMIC RESTRUCTURING OF LONDON

In 1992, Greater London had a population of 6.9 million (11.9 per cent of the total UK population), which is larger than those of many nations, including Finland, Singapore and Ireland. So too is its economy. With a GDP of $122 billion in 1989, the London economy is about the same size as the economies of Russia ($125b), Thailand ($117b), Hong Kong ($111b) or Finland ($105b), and noticeably larger than those of Greece ($80b), Portugal ($80b), Singapore ($52b) and Ireland ($50b). In fact, if London were a member of the EU in its own right, four other member states (Greece, Portugal, Ireland and Luxemburg) would have smaller economies (McWilliams 1993).

The London economy, moreover, is of greater importance than its sheer size indicates. Since it possesses many of the control and command functions of the world economy, it is from London that decisions are taken which determine the fate of vast areas of the world. Examining the distribution of the headquarters of the 500 largest transnational corporations (TNCs) in the world in 1984, Feagin and Smith (1987) find that ten cities housed just over 50 per cent of the world's top 500, New York having the largest concentration with 59 headquarters and London having the second highest concentration with 37.[1] Indeed, this city is at the pinnacle of one of the most centralized corporate systems of any nation in Europe, possessing 271 (54 per cent) of the head offices of the UK's largest 500 companies (Marshall and Raybould 1993).

It is a city, furthermore, which has witnessed a rapid and radical restructuring of its economy. Deindustrialization has proceeded faster and further in London than both the UK in general and other global cities. In 1971, London's economy employed over a million manufacturing workers, who produced 23.4 per cent of the capital's GDP. By 1991, only 358200 remained (11.0 per cent of London's workforce, compared with 21.2 per cent nationally) and manufacturing produced just 12.7 per cent of London's GDP as against 20 per cent nationally (Census of Employment 1991, McWilliams 1993). Indeed, amongst other global cities, only New York shares a similar rate of manufacturing decline and such a low share of total employment in manufacturing (Kennedy 1991).

So, compared with both the national economy and other global cities, London is essentially a service-based economy. By 1991, and as Table 14.1 reveals, the service sector employed 84.1 per cent of the London workforce, compared with just 71 per cent nationally, and this sector produced 81.5 per cent of its GDP as against 66 per cent nationally (McWilliams 1993). This domination of the London economy by services, however, is not solely due to the preponderance of producer services. Although the UK producer services sector is concentrated in London, especially the larger producer service firms and their control functions (see chapter 3), such services employ just 15.5 per cent of the London workforce.

Consumer services, meanwhile, employ 38.9 per cent of London's employees. Indeed, between 1981 and 1991, the vast majority of the employment growth in London came from net job gains in this sector. As Table 14.1 displays, producer and consumer services were the only two sectors of the London economy to witness a net increase in jobs during the 1980s, with the consumer services sector dominating this expansion, creating nearly twice as many jobs as producer services (94100 compared with 49100). Indeed, many of the fastest-growing service industries in the London economy have been consumer service industries, including two of the top three (see Table 14.2). Are these consumer service jobs, however, simply a by-product of the wealth generated in other sectors of the London economy? Or are they, like second-tier cities and rural areas, functioning in a manner which increases net income for this city? To answer this, one needs to examine the nature of trade in London.

To examine what London is trading with the rest of the world, McWilliams (1993) examines the final demand of Londoners and the level of production of each good and service in the city so as to identify the shortfall or surplus in each good and service. This reveals the extent to which London is an open economy (see Table 14.3). Some 52 per cent of its total turnover is exported and the magnitude of its imports is equivalent to 42 per cent of its total production. Breaking down these imports and exports by sector, he finds that manufactured products make up the largest single category of imports, accounting for 67 per cent of the total, although primary products are also heavily imported (see Table 14.3). Services, meanwhile, are the major export, constituting 63.2 per cent of all exports.

What services, however, are being exported? Are they, as many assume, producer services in general, and financial services in particular, or do consumer services also contribute to external

Table 14.1 Sectoral distribution of employees in employment in Greater London, 1981–91

	1981			1991			% change 1981–1991	
	No. (000s)	%	LQ	No. (000s)	%	LQ	London	UK
Non-services	903.3	25.4	0.64	518.6	15.9	0.55	−42.6	−19.6
Services	2656.3	74.6	1.21	2736.1	84.1	1.18	3.0	18.6
Consumer	1171.2	32.9	1.01	1265.3	38.9	0.97	8.0	26.0
Producer	456.8	12.8	1.49	505.9	15.5	1.44	10.7	28.4
Mixed	731.0	20.5	1.60	717.0	22.0	1.61	−1.9	8.9
Public	297.3	8.3	1.17	247.9	7.6	1.21	−16.6	−9.6

Source: Census of Employment 1991

Table 14.2 Ten fastest-growing service industries in London, 1981–91
(£m)

Industry	% increase in jobs
Medical practices	127.6
Hiring: office machinery/furniture	118.5
Tourist offices/other community services	114.5
Activities auxiliary to banking/finance	110.5
Hiring: other movables	77.7
Business services	66.6
Legal services	57.8
Authors, musicians, composers, etc.	57.4
Commission agents	55.8
Hiring: transport equipment	55.4

Source: Census of Employment 1991

Table 14.3 Production, imports and exports for London, 1989 (£m)

Industry	Turnover	Imports	Exports
Primary	3588.80	6704.62	411.80
Manufacturing	24521.38	37317.02	15669.38
Construction	9047.75	1232.78	2062.02
Distribution, hotels, etc.	30780.67	3175.18	18079.94
Banking, finance, etc.	58312.85	6719.04	31436.70
Public administration	3856.32	—	—
Total	130107.77	55148.64	67659.84

Source: McWilliams (1993: Table 4.3)

income generation? Although Table 14.3 does not differentiate completely between producer and consumer services, it does reveal that the producer-oriented sector of banking, finance, etc. exported £31.4 billion worth of services (63 per cent of total service exports and 46.5 per cent of all exports), whilst the consumer-oriented sector of distribution, hotels, etc. has exports worth £18.1 billion (37 per cent of all service exports and 26.7 per cent of total exports). Consequently, although much attention has been lavished on London's producer services, especially its financial service industries (Daniels 1986, 1995a, Gentle 1993, Leyshon et al. 1993, Marshall et al. 1992, Thrift 1990, Thrift and Leyshon 1992), these are not the only service activities contributing to external income generation. Indeed, the foreign export earnings of London's financial services, at £6.2 billion in 1989, comprise just

21.7 per cent of service-sector export earnings and 9.2 per cent of total exports (derived from McWilliams 1993). It is clear, therefore, that in London, as in the other localities which have been studied, consumer-oriented service industries display an external income-generating function.

Indeed, consumer services in London are higher up the sectoral hierarchy in terms of their proclivity to export than many may have so far assumed. Manufacturing, exporting 63.9 per cent of total turnover, is at the top of this hierarchy, followed by distribution, hotels, etc., whose exports equal 58.7 per cent of its turnover, and only then banking, finance, etc. at 53.9 per cent. The construction industry, meanwhile, exporting 22.7 per cent of turnover, and primary industry exports at 11.5 per cent are at the bottom of the hierarchy (derived from Table 14.3). Consequently, consumer-oriented service industries generate greater proportions of their income externally than producer-oriented services.

As argued in previous chapters, however, ranking sectors by their export propensity is neither a wholly accurate nor an appropriate measuring rod of the contribution of a sector to economic development. More salient is its contribution to increasing net income (exports minus imports). Examining this, Table 14.3 reveals that the producer-oriented banking, finance etc. industries generate a net income for London of £24.7 billion whilst the consumer-oriented distribution, hotels, etc. industries produced £14.9 billion in net income (or 37 per cent of total service-sector net income). The consumer-oriented industries, therefore, make larger contributions when their net income impacts are examined than when solely their exports are considered because they are imported to a lesser extent and thus leakage of income is lower. In consequence, the income generated by London's service sector can no longer be assumed to derive mostly from producer services. Consumer services are also important.

THE ROLE AND CHARACTER OF CONSUMER SERVICES IN A GLOBAL CITY

According to Christaller's central place theory, London as a first-order centre at the top of the national hierarchy should be over-represented by consumer services. As Table 14.1 shows, however, it is not. To a large extent, this can be explained by the size of the commuting population, in that such workers acquire some of their routine consumer services (for example food, restaurants, welfare services) in outlying areas. This under-representation of consumer services in London, however, should not be taken to mean that these industries result in a net deficit for the London economy. As indicated, this is not the case. This is because consumer services do not simply meet local demand, as insinuated by Persky and Wiewel (1994) when they argue that the expansion of consumer services implies a 'growing localness of the global city'.

To display both the role and character of consumer services in this global city, a wide range of consumer services could be examined. Mohan (1988a and b, 1991), for example, shows that private health care is concentrated in London and that its specialized forms make major contributions to the London economy. Universities, moreover, could be shown to be not only relatively concentrated in London but major generators of net income (see chapter 7). In 1992, for example, the University of London spent £1.5 billion, which is some 2 per cent of London's GDP, and received 25.3 per cent of its income from overseas students (Goldsmiths' College 1995). Here, however, and to illustrate the contributions of consumer services in this global city, two consumer services in particular will be briefly analysed: tourism and the cultural industries. This will reveal that although London is characterized by an under-representation of the consumer services sector as a whole, this global city not only generates considerable net income from these industries, like many second-tier cities and rural areas, but also, and in contrast to these other localities, harbours this sector's headquarters as well as higher-order consumer services, thus reinforcing in the context of consumer services the uneven patterns of development already identified in other producer services and manufacturing (for example Daniels 1993, Dicken 1992, Marshall and Wood 1995).

London's tourist industry

To see that consumer services in global cities are not dependent activities, one has only to examine the tourist industry in London. As Table 14.4 indicates, the 20.1 million British and foreign tourists visiting London in 1994 contributed £6285 million to this city, which the Foreign and Commonwealth Office (1995) suggests supports 200000 jobs. This means that London's tourist industry generates about the same level of export earnings as its financial

Table 14.4 Tourism visits and expenditure in London, 1992–6

	1992	1993	1994	1995 (est.)	1996 (est.)
Visits (m)					
Domestic	7.0	7.2	8.6	8.8	8.8
Overseas	10.0	10.2	11.5	12.9	13.5
Total	17.0	17.4	20.1	21.7	22.3
Expenditure (£m)					
Domestic	640	875	1005	1160	1160
Overseas	4150	4850	5280	6315	6825
Total	4790	5725	6285	7475	7985
London's share of overseas tourism in UK (%)					
Visits	53.8	52.8	54.5	54.5	54.5
Nights	36.9	36.5	38.8	38.8	38.8
Expenditure	53.2	52.4	53.8	53.8	53.8

Source: London Tourist Board and Convention Bureau (1996)

services industry, even though these tourism figures are an under-estimate since they exclude the spending of the 70 million tourists who annually make day visits to London (London Tourist Board and Convention Bureau 1995). Tourism, unlike financial services, is also a major growth industry. Between 1992 and 1996, for example, tourist visits to London rose by 31.1 per cent, and expenditure by 66.7 per cent (see Table 14.4).

Tourism, moreover, reinforces rather than mitigates the spatial inequalities between the core and periphery produced by other sectors of the economy (Christaller 1963, Peters 1969). In 1994, 54.5 per cent of overseas tourists visited London, 53.8 per cent of their expenditure was in this global city (see Table 14.4), and foreign visitors alone account for 8.8 per cent of all consumer expenditure in London (McWilliams 1993). The outcome is that tourism in London generates additional annual income equivalent to £911 per head of population compared with just £268 in Sheffield, which is well over three times higher (London Tourist Board and Convention Bureau 1996, Destination Sheffield Visitor and Conference Bureau 1994).

Furthermore, there is little evidence of any significant decon-centration over time. This is because many of the expanding and lucrative forms of tourism remain essentially capital-city-focused activities. As revealed in chapter 5, international conference and exhibition tourism, international cultural tourism (see below) and business tourism all reinforce London's dominant position in UK tourism, the latter contributing some £1.7 billion in visitor expenditure in 1994 to the London economy (London Tourist Board and Convention Bureau 1995). So, although the British Tourist Authority (BTA) have taken steps to decant tourists to other provincial destinations by trying to ensure that at least 60 per cent of overseas visitor nights are spent out of London (House of Commons Employment Committee 1990), the results have been limited. London's share of total expenditure by overseas tourists visiting the UK dropped only from 57.8 per cent in 1980 to 53.8 per cent in 1994 (London Tourist Board and Convention Bureau 1995). Indeed, during the 1990s, it has slightly increased its market share (see Table 14.4).

Consequently, the tourist industry remains a consumer service which reinforces the dominant position of London in the national economy. Indeed, many of the major tourist facilities remain London-based. The six most heavily visited museums and galleries in the UK, for example, are all located in London: the British Museum, National Gallery, Tate Gallery, Science Museum, Victoria and Albert Museum and Natural History Museum. Neither should the much-heralded relocation of exhibits from the Royal Armouries to Leeds and from the Victoria and Albert Museum to Sheffield be over-emphasized as evidence of decentralization. This is simply the relocation of surplus artifacts for which no room is available in London. It is not the relocation of the core facility. In this sense, such relocations are akin to producer services decanting lower-order back-office functions to peripheral regions. Nor is there much evidence of major new facilities and events being relocated out of London. The new national stadium, for instance, is again likely to be built in London at Wembley, and the Millennium Exhibition, expected to attract 30 million tourists, has been awarded to Greenwich in London, yet further concentrating tourist expenditure in the capital.

The effect is that despite some shifts in the tourist products available, increased competition from second-tier cities, and some lip-service paid to a spatial policy towards tourism, London remains a dominant tourist destination within the UK. It is also the home of the control and command functions, with most of the major corporations in the increasingly oligopolistic tourist industries of hotels, travel agencies and airlines having their corporate headquarters in this global city. So, as with producer services and manufacturing, despite some limited deconcentration of back-office functions and lower-order services to outlying areas in recent years, the control and command functions remain concentrated in London. This consumer services industry, therefore, has a similar uneven distribution to other sectors of the economy, which serves merely to reinforce the position of London in the national economy. It is a similar case, as will now be seen, with the cultural industries sector.

The cultural industries in London

Although much time and effort has been spent examining how the cultural industries might help regenerate sub-global cities and other localities in recent years (see chapter 9), all the evidence suggests that they remain firmly embedded in London. Whilst 14.6 per cent of all British employees in employment worked in London in 1993, 39.9 per cent of employees in motion picture and video activities; 52.8 per cent of those in radio and television activities; 51.4 per cent of those in live theatrical presentations; 29.1 per cent of those involved in the operation of arts facilities; and 74.2 per cent of those in news agency activities were working in London.

It is not only the concentration of cultural industry employment in this global city which is important, but also the fact that it houses most of the control and command functions of the major companies involved in cultural industry production and many of the principal outlets for cultural-oriented consumption. To show this, the structure of several sub-sectors of the production- and consumption-oriented cultural industries is examined.

Take, for example, the printing and publishing industry, which embraces newspaper, periodical and book publishing, as well as other types of printing and publishing such as promotional and corporate publishing, including related activities such as printing, bookbinding and the manufacture of paper goods machinery and printing ink. According to Comedia (1992), the publishing industry as a whole in London comprised 1500 businesses in 1989, employing nearly 90000 people and generating a turnover of £2720 million, and is by far the largest employer of all London's cultural industries. This is an industry, moreover, which is heavily concentrated in London. In newspaper publishing, for instance, eleven out of twelve of the national dailies are based in London, as are all eleven of the national Sundays (Urban Cultures Ltd 1994). Moreover, it is an oligopolistic industry. Four newspaper groups produced 87.3 per cent of the national daily and Sunday newspapers sold each week in 1994, including News International (34.2 per cent of the market); Mirror Group Newspapers Ltd (27.1 per cent), United Newspapers (13.3 per cent) and Daily Mail and General Trust PLC (12.7 per cent), all of which are based in London.

In periodical publishing, it is a similar picture. The industry is dominated by large companies (though not to the extent of the newspaper industry) which primarily publish business and consumer titles. The Periodical Publishers Association has 180 members, who publish between them 1400 titles generating 80 per cent of the sector's revenue. The majority of these companies are either based in London or have some kind of base in the capital. As with many producer services, furthermore, although there has been some outward movement during the past decade or so, this has been a concentrated decentralization either to outer boroughs of London or to locations in the rest of the South-East (Urban Cultures Ltd 1994). Again, London is unquestionably at the core of book publishing. Other large cities such as Birmingham, Manchester and Glasgow have only a handful of publishers. Perhaps the most significant competition comes from Oxford and Cambridge, traditionally the home of academic publishing but attracting a wider range over the past two decades or so (Urban Cultures 1994).

It is not only in the realm of printing and publishing that London is the dominant location for the cultural industries both in terms of employment and as the hub of the control and command functions. It is a similar case in many other cultural industries. Take, for instance, the music industry. Comedia (1992) estimate that in 1989/90, there were 2650 businesses and organizations working in the music industry in London, employing just under 30000 people and with a turnover of £395 million, a relatively high proportion of which is export earnings. As indicated in chapter 9, however, five major multinationals dominate this industry (EMI, Sony, Polygram, BMG and Warner Brothers/Time Warner), all of which have their headquarters in central London.[2] The smaller companies, meanwhile, only survive by entering relationships of collaborative competition with these majors, using the 'indies' as their artists and repertoires (A & R) arms. This has led to hybrid organizations known as 'mandies' (major + indy), which search out new talent on behalf of the parent company (Urban Cultures Ltd 1994). London, therefore, remains at the heart of this industry on a national level with 'subsidiary' companies in the provinces acting as 'talent scouts' for the London-based oligopolies.

Besides these production-oriented cultural industries many consumption-oriented cultural industries are concentrated in London. An example is the performing arts. By 1993, 7000 of the 13600 employed in Britain in live theatrical presentations were working in London (51.6 per cent). This is an industry, moreover, which generates considerable amounts of external income. Comedia (1992) estimate that 11.7 per cent of the income generated by the performing arts in London comes from overseas visitors, whilst Myerscough (1988) asserts that 40 per cent of those attending London theatres and 8 per cent of those attending concerts are tourists.

It is a similar story with museums and galleries. As stated above, all of the six most-visited museums and galleries in the UK are located in London, and this consumer-oriented cultural industry again has the ability to be a major generator of external income. Myerscough (1988) finds that tourists account for 44 per cent of those attending such facilities and foreign tourists alone for 31 per cent of total attendance in London. On the basis of research conducted in 1986 in four major London galleries (the Royal Academy of Arts, the Barbican Gallery, the Hayward Gallery and the Tate Gallery), Myerscough (1988) finds that for every 100000 visitors to a temporary loan exhibition in London, some £6.03 million was injected into the London economy. This is net of deadweight expenditures which would have occurred anyway irrespective of the exhibition. Such evidence thus reveals that far from being a cost to the public purse, temporary loan exhibitions represent a major contribution to the London economy.

In sum, in both the production- and consumption-oriented cultural industries, London dominates the national economy. Cultural industry jobs are concentrated in this city, and so too are the control and command functions and higher-order consumer-oriented cultural industries. The result, as Kennedy (1991) reports, is that the cultural industries have a turnover of some £7500 million per annum in London alone (which is again around the same level as the whole of the financial services sector) and they directly employ 214500 people, which represents around 6 per cent of total employment in London. Together, moreover, they grew by about 20 per cent during the 1980s.

This sub-sector of consumer services, however, does not contribute to the London economy solely in terms of its direct revenue-generating effects. It also plays a pivotal role in establishing a positive image for this global city, which, as outlined in chapter 9, is a crucial consideration in the global marketing of the metropolis so as to attract inward investment in other sectors of the economy as well as improve the quality of life of existing and potential residents. Thus, and as with many other consumer services, the preponderance of such services in London creates a milieu which facilitates the retention and attraction of both producer services and manufacturing (McKellar 1988, Montgomery 1995a).

However, this comparison of London's cultural industry sector with the rest of the UK, although necessary to show the dominance of London in the national context, is perhaps insufficient to demonstrate the pressures on London's cultural industries sector. As in manufacturing and producer services, competition takes place at a strictly hierarchical level. Whilst the multiplicity of second-tier cities compete with each other for the lower-order functions and facilities in the cultural industries, London is more involved in a fierce international competition between global and capital cities, especially for higher-order functions in this sector, as currently witnessed in the battle between Paris, Milan and London to become international centres of fashion and culture. When examining London, therefore, it is perhaps more important to evaluate its position in relation to other global cities.

To do this, Comedia (1992) examine each sub-sector of the cultural industries in terms of its relative competitiveness with other global cities at each stage in the production chain from the generation of ideas to the point at which the product is consumed by an audience. Five stages are thus examined: beginnings (ideas generation, as displayed by copyrights, trademarks, patents); production (the transformation of the ideas into marketable products, measured by the level and quality of impresarios, managers, producers, editors and engineers as well as suppliers and makers of equipment such as in film and design, studios and frame-makers); circulation (the quality of agents and agencies, distributors and wholesalers (in film and publishing), packagers and assemblers); delivery mechanisms (the platforms which allow cultural products

Table 14.5 London's performance relative to other world cities in the cultural production chain

	Beginnings	Production	Circulation	Delivery mechanism	Audience reception
Performing arts	4	3	3	5	3
Music	4	4	4	3	4
Film and television	4	2	3	2	4
Literature	3	4	3	3	3
Visual arts	4	2	3	3	3
Design	4	2	3	3	3
Architecture	4	1	2	n/a	2

1 = very weak performance
5 = very strong performance

Source: Comedia (1992)

to be consumed, such as cinemas, theatres, bookshops, TV channels, screens, magazines, museums, record shops); and audiences and reception (which concerns the public and critics). Table 14.5 displays London's performance at each of these stages in relation to other global cities.

From this, it is evident that London is very strong at the front end of the creative process but weaker in the latter stages. The problem, therefore, is that the latter higher value-adding stages of the cultural production process may well be lost to other global cities. Greater support is therefore required for turning cultural creation into products with avenues for their consumption. Although London remains extremely accomplished in promoting tried and tested cultural products, for example, the outlets for the promotion of innovative new-wave products have received uneven support and success so far as investment is concerned. As a result, the city has seen a weakening of the outlets through which new ideas are fed into the cultural production system. As Kennedy (1991: 44) concludes, 'Other world cities appear to subscribe to the view that *culture creates wealth*, whereas in London the prevalent view is that *wealth creates culture*' (his italics). Although this is perhaps an exaggeration, it is nonetheless true that without investment, the current position of London as a global centre of cultural production and consumption would come under increasing threat.

If this were to occur, it would have important knock-on effects both for other economic sectors in London and for the national economy. Investment is thus required just as much in the cultural industries in particular, and consumer services in general, as it is in other sectors of the economy. With this in mind, we can consider the implications for policy of the above analysis.

POLICY IMPLICATIONS

The London economy, therefore, has witnessed severe deindustrialization and a rapid tertiarization of its economy. Although the conventional approach has been to emphasize the contributions of the producer services in general, and financial services more particularly, to the vitality of the London economy, given its position as a global city possessing the command functions of the world economy, this chapter has argued that the contribution of consumer services cannot be ignored or relegated to a dependency status. It is a major generator of external income for the London economy. Indeed, when the specific consumer service industries of tourism and the cultural industries sector are reviewed, each is revealed to contribute as much in foreign export earnings to the London economy as the whole of the financial services sector. The contribution of consumer services, moreover, is not solely in terms of generating external income. They also improve the competitiveness of the other sectors of the London economy by enhancing the quality of life and making London an attractive location for businesses and the new service class. For that reason, they are important not only as basic sectors of the London economy but also through the way in which they inter-relate with the other sectors of the economy. For this global city to continue to prosper, therefore, greater emphasis will be required on consumer services.

In this regard, several key questions will require consideration. How do consumer services need to develop to make London competitive with other global cities? Should there be a focus upon developing such consumer services in the London economy if this is at the expense of the rest of the UK economy? Or are there trickle-down effects? If so, how can these be improved? Is the taxation system the most appropriate method for redistributing this wealth? Alternatively, should specific consumer service sectors or key consumer service industries be decentralized so as to encourage growth poles and/or specialized consumer service industrial districts elsewhere in the UK? Future research will have to address such questions since until now, there has been little, if any, thought given to the role of consumer services in the planning of the UK spatial economy.

Besides such spatial distribution questions arising out of the recognition that consumer services can act as motors for economic development, there are also questions concerning social justice which ensue from this analysis. Sassen (1988, 1991), for example, has argued that there is an increasing polarization in occupational and income structures in global cities since the shift from manufacturing to services has resulted in a decline in middle-income jobs and an increase in high- and low-paying jobs at each end of the occupational and income hierarchy. The outcome is a 'dual city'. In this view, therefore, the shift towards services in global cities results in social polarization. Hamnett (1994a and b), however, has argued that this is particular to New York and Los Angeles and asserts that London has witnessed more a professionalization of the occupational structure,[3] whilst Williams and Windebank (1995a) have shown that income polarization at a household level is less marked in London's service-dominated economy than elsewhere in Britain. This calls into question whether social polarization in London is as closely correlated with the shift to services as Sassen suggests. Future research will need to investigate this in greater depth, especially the common assumption that consumer services only create disproportionate numbers of low-paid and low-skilled jobs.[4] However, and whatever the finding on the role of services in social polarization, the important finding to arise out of this chapter is that consumer services, far from being dependent activities, generate net income in this global city and act as the control and command points for both the rest of the national economy and much of the global economic system.

15

CONCLUSIONS
Rethinking consumer services and local economic development

INTRODUCTION

Grounded in economic base theory, which assumes that any economy must earn external income in order to grow, conventional economic development theory and practice discriminate between 'basic' activities, which generate such external income and thus act as engines of growth, and 'dependent' activities, which merely circulate income within a particular economy. In consequence, the principal focus of local economic development has been on cultivating the basic sector, traditionally believed to be composed of primary and secondary activities. A major achievement of the vast array of service-sector studies conducted over the past decade or so, however, has been to redefine producer services as basic-sector activities and thus motors of local economic development. The old 'manufacturing as engine of growth' and 'services as dependent activity' dualism has been largely replaced by a 'manufacturing and producer services as motors of development' versus 'consumer services as parasitic' dichotomy.

Remaining largely unquestioned, however, have been first, the view of the economy as consisting of a basic and dependent sector and, second, the perception that an economy can be structured into a hierarchy of sectors ordered according to their ability to export. Focusing upon the role of consumer services in local economic development, the main aim of this book has been to evaluate critically both of these conceptualizations. Here, therefore, and by way of conclusion, the findings from the studies of specific consumer service industries and the function of consumer services in particular localities are drawn together to recast the role of consumer services in local economic revitalization as well as the

235

theory and practice of local economic development themselves. To do this, first, the evidence that consumer services are not dependent activities will be reviewed, second, the notion of a sectoral hierarchy based upon export propensity will be evaluated, and finally, the implications for local economic development will be explored.

CONSUMER SERVICES AS DEPENDENT ACTIVITIES: A CRITICAL EVALUATION

By revealing that consumer services, as the only sector remaining entrenched in the dependent category, do in fact fulfil an external income-generating function, this book has raised serious doubts about the validity of continuing to employ the basic/dependent sector dichotomy in economic development theory and practice at all. Up to now, the problem has been that individual consumer service industries have been studied in isolation, which means that although specific consumer services have been revealed to be basic industries, the dualism itself has gone unchallenged. This book, however, by treating the consumer services sector as a whole and examining a wide range of consumer services, has brought this basic/dependent sector dualism starkly into question.

Examining tourism (chapter 5), sport (chapter 6), universities (chapter 7), retailing (chapter 8) and the cultural industries (chapter 9), the book has uncovered how they all function as basic activities, albeit by importing consumers rather than exporting products. Although some consumer services (i.e. universities, tourism) are more external income-oriented than others (i.e. retailing, sport and the cultural industries), all earn a proportion of their income externally.[1]

Indeed, the locality studies in Part III indicate the nature and extent to which consumer services are generating external income. In London, for example, and as chapter 14 displayed, not only do consumer-oriented services generate greater proportions of their total income externally than producer-oriented services, but the external income induced by either the tourist industry or the cultural industries sector is equal to that of the financial services sector. In sub-global and regional cities such as Glasgow (chapter 10), Sheffield (chapter 11) and Leeds (chapter 12), meanwhile, consumer services have been shown to be not only one of the principal sources of job growth and amongst the largest employers in these cities but also increasingly functioning as basic industries as well as catalysts for expansion in other sectors. Even in rural localities such as the Fens, where consumer services might be least expected to play a significant role in local economic development, this sector has been revealed to contribute to external income generation (see chapter 13). The finding is that consumer services as a whole should be redefined as basic-sector activities, similar to manufacturing and producer services.

This recognition of the external income-generating propensity of consumer services leads on to a radical rethinking of the geography of consumer services. Christaller's central place theory, for example, which conceptualizes such services as simply meeting local demand and orders them into a hierarchy of centres based on threshold populations, is both far from the reality and fails to recognize the complexity of how consumer services are unevenly distributed. Not only is this sector under-represented in both the global city of London (see chapter 14) and some second-tier sub-global and regional cities such as Leeds (see chapter 12), but their spatial distribution, as has been shown, varies according to sub-sector considered, ownership status and firm size, to name but a few variables. There is also evidence of local and regional specialization in consumer services. Glasgow, for example, appears to be specializing in the cultural industries and Sheffield in sport, whilst localities such as St Andrews, Cambridge and Oxford specialize in higher education. So too is there evidence of new 'industrial districts' being forged based on consumer services, which replicate many of the characteristics of earlier industrial districts founded on manufacturing (see chapter 11).

Moreover, and as in the case of producer services and manufacturing, evidence has been provided of an increasing spatial division of consumer services based on a separation of policy-making from service provision (for example in the cultural industries, tourism and retailing). Frequently, this takes the form of a periphery/core pattern, with the control and command functions and higher-order consumer services concentrated in London and routine lower-order functions more evenly distributed. As with producer services, furthermore, although some limited decentralization of these policy-making functions has sometimes occurred (for example in tourism and publishing), this has taken the form of a concentrated decentralization.[2] Consequently, if any trend is discernible, it is that the uneven distribution of consumer services reinforces, rather than mitigates, many of the spatial inequalities produced by the other sectors of the economy.

Neither is this likely to change in the near future or even beyond. The concentration of capital in consumer service industries (for example retailing, tourism and the cultural industries) is strengthening this spatial division of consumer services, as is the separation of the consumer from the producer because of the advent of new innovations such as telecommunications, as Marshall and Richardson (1996) display in the context of telemediated retail banking. This recognition of the external income-generating propensity of consumer services, therefore, means that rather than a reliance on out-dated theorizations of the geography of consumer services, a profound reassessment is required. This book has shown that the developments in the study of producer services in particular, and manufacturing more generally, seem to represent a useful starting point for such a reconfiguration.

Indeed, this new geography of consumer services will need to take full account of the state of dynamic flux in this sector resulting from the interplay between global trends, local conditions and institutional policy regimes. The appreciation that consumer services generate external income, for example, has led many economic development agencies to try actively to reshape the geography of consumer services in ways favourable to their localities. Glasgow, for example, has reinvented itself as a 'city of culture' whilst Sheffield has placed much emphasis on the sport, cultural industry and retail sectors in its economic development strategy (see chapters 10 and 11). It is more than mere coincidence, however, that the localities which have most heavily turned towards consumer services as their saviour are those which felt that they had little other option open to them following the severe restructuring of their manufacturing base. Both Glasgow and Sheffield, as shown, adopted consumer services at the heart of their regeneration strategies following a period of widespread deindustrialization and because of a poor performance in attracting producer services. However, in other cities more successful at retaining and attracting other sectors, such as Leeds (see chapter 12) and London (see chapter 14), consumer services have received much less attention. Instead, at best, they have been seen in local economic policy as making a supportive contribution by bolstering the attractiveness of the city to producer service and manufacturing companies. Consumer services, therefore, although increasingly recognized as having an external income-generating role to play in economic development, are only embraced as a lead sector by localities when other possibilities have been ruled out by economic development agencies.

Consequently, this book has revealed that despite some acknowledgement that consumer services are not dependent activities, there remains at the heart of economic development thinking and practice a 'hierarchy of sectors' approach which prioritizes sectors according to their perceived ability to generate external income and relegates consumer services to the lowest rank in this hierarchy. Manufacturing is seen to be at the top of this hierarchy, followed by producer services and only then consumer services. After a study of the contributions of consumer services to economic development, however, serious misgivings have been voiced about this hierarchical approach. Commencing in chapter 4, and further illustrated and explained throughout the sub-sector and locality studies in Parts II and III, it has been shown that to understand fully the contribution of consumer services (and other sectors) to local economies, one has to go beyond an appraisal of their basic-sector characteristics alone.

FROM HIERARCHIES TO HOLISM IN ECONOMIC DEVELOPMENT THEORY AND PRACTICE

To focus upon an industry's basic-sector characteristics is to assume that economic growth is merely a product of the level of external income-generating activity in a locality. However, throughout this book it has been revealed that economic growth is not so strongly correlated with income-inducing activity as many might have previously assumed. This is because for an economy to grow, it is not a rise in external income alone but, rather, an increase in net income which is required. Net income, as has been illustrated, is determined by total external income, times a multiplier (which is higher the more self-reliant the economy), minus total external spending. The growth of a local economy is therefore just as dependent on preventing leakage of income out of its area as it is upon external income generation.

In recognition of this, there is a need to move beyond the sectoral-hierarchy view based on export-orientation and towards a more holistic understanding of the economic development process. This holistic approach argues that to comprehend the contributions of an activity to economic development, one needs to examine both its propensity to generate external income and its leakage-preventing abilities. The problem, however, and as the preceding chapters have revealed, is that this is not currently occurring. In chapter 4, the idea was introduced that there has been little, if any, appreciation either that locally oriented activity is as important to local economic development as outward-oriented activity or, more specifically, that local consumer services are as vital as basic consumer services. All the chapters on both individual consumer services in Part II and particular localities in Part III have reinforced this perception.

Take, for example, tourism. Chapter 5 showed that despite the massive surge in overseas tourism to the UK, tourism policy retains a focus on accommodating ever more visitors and little, if any, emphasis on leakage prevention. Yet the UK economy has frequently made a net loss on tourism because of the equally rapid growth in UK tourists visiting foreign destinations. Recognition of the fact that leakage prevention is just as important as external income generation could thus result in major policy changes in tourism provision and policy. The aim would be to give equal emphasis to retaining UK tourists within the nation by providing facilities that they demand and to focusing upon the desires of foreign tourists.

Similarly, cities such as Glasgow, as shown in chapter 10, have emphasized the enticement of external visitors rather than providing facilities for local cultural tourists. Given that 90 per cent of resident spending on local cultural provision would occur outside the locality if facilities were not available within the area (Myerscough 1988), a more inward-looking policy would retain this spending by the local population rather than allow it to be dispersed. For this to be achieved, however, and as much of the cultural industry literature reported in chapter 9 recognizes, the need is for a radical change in the nature of cultural industry provision. Facilities provided to attract external cultural tourists, such as 'flagship' projects, are not necessarily the same as those required to retain the expenditure of the local population and create the local synergies required for successful regeneration.

The justification for both greater emphasis on leakage prevention and thus the inter-dependencies between all sectors of the local economy, nevertheless, stretches far beyond this purely economic rationale. There is also a strong environmental argument for

CONCLUSIONS

organizing both production and consumption on a more locally oriented basis, not least because of the resource conservation aspects which result from such a policy (for example reduced transport costs). Indeed, the holistic approach to local economic development recommended here is very much akin to the self-reliance strategies advocated by many of those who discuss sustainable development from a 'deep ecology' perspective (for example Dauncey 1988, Dobson 1993, Ekins and Max-Neef 1992, Robertson 1989). At the heart of this self-reliant approach is the principle of using local products and services whenever possible and importing only when these products and services cannot be made available locally.[3] The environmental rationale for the localization of economies, however, is itself the subject of another book, and can only be touched upon here.

Nevertheless, given the rationale for promoting this more 'holistic' approach to economic development, the key issue that remains is how such an approach can be implemented. What are the principles which a locality seeking to implement such a strategy should follow?

IMPLICATIONS FOR ECONOMIC DEVELOPMENT

To execute this holistic approach, Part II highlighted the policy shifts required in each of the individual sub-sectors and Part III considered the policy implications in the context of the localities studied. Here, the common strands identified in each of these chapters are woven together to form a web of principles to guide local economic development. This is composed of four principal inter-locking goals: encouraging local ownership; increasing import substitution; improving the local control of money; and localizing work to meet local demand. All are inter-related, but are here considered in turn. The suggestion, moreover, is that the consumer services sector is an appropriate place to start because it is perhaps easier to localize than manufacturing. This is not because it is already more locally oriented. Rather, it is because there is perhaps more consensus that consumer services should be localized activities amongst the various stakeholders engaged in formulating local economic policy.

Encouraging local ownership

In the export-oriented paradigm, simply encouraging activity which generates external income is sufficient reason for supporting an industry. The result is that many local economies encourage externally owned large corporations to relocate in their area with little regard for their overall impacts on the local economy. Given that these corporations frequently source inputs from outside the area, the only benefit for the locality can be the wages and the resulting multiplier effects resulting from employees' expenditure (see, for example, Foley *et al.* 1996).

In this holistic approach, however, where emphasis is placed as much on leakage prevention as external income generation, locally owned businesses are encouraged. This is because these firms are asserted to be more likely to retain income within the local economy by recirculating profits and assets locally, and to purchase local services and products, as displayed in the Fens (chapter 13). They also increase jobs at a faster rate than externally controlled firms and tend not to cut back jobs to the same extent as their non-local counterparts during downturns (Bluestone and Harrison 1982). Therefore, developing locally owned small retailers, for instance, is seen as preferable to the imposition of food superstores owned by the major retail corporations, which source their goods externally, and locally owned hotels and guest houses are preferred to hotel chains.

Increasing import substitution

Running alongside this principle of encouraging local ownership is that of facilitating the local production of goods and services which are currently imported so as to decrease the reliance of a locality on imports. This reduces leakages and increases the local economic multiplier effects. How, nevertheless, can this shift from an outward-looking orientation to an inward-looking trajectory be achieved in economic development strategies? The first, and most obvious method is for local economic development agencies to identify the types of imported products and services that can be replaced by local production, inform local firms of these opportunities and encourage them to compete for this business and provide the necessary technical and marketing assistance. Some of the mechanisms to do this already exist so far as intermediate demand is concerned, such as Better Business Opportunities' (formerly Better Made in Britain's) REGAIN initiative (see Williams 1994a), which seeks to identify on a sector-by-sector basis the current imports into a locality and, following this, to target the potential local suppliers of these goods and services and put them in touch with each other.

Second, and in terms of final rather than intermediate demand, 'use it or lose it' campaigns to encourage consumers to buy more local products are one possible approach. This method, however, and as displayed in chapter 13, is not only largely unsuccessful but even if people do buy locally, the business may often be owned externally and/or the goods and services sourced from external suppliers, so a share of the profits and income will quickly leak out of the area. It will not be recirculated within the economy. The result is a low multiplier effect. Such problems can be overcome by drawing upon innovative local currency systems both to retain and to circulate money within the local economy.

Encouraging local control of money

One of the principal problems at present in local economies, as Jacobs (1969) recognized as early as the 1960s, is that income leaks out of localities since it is not locally controlled. There are, however, several ways to overcome this. First, cities could invest portions of the municipal pension funds in local projects, so that fiscal resources which might otherwise escape the locality can be retained (Keating 1991, Leatherwood 1983, Rifkin and Barber 1978). Second, credit unions could be used to invest money locally (Leyshon and Thrift 1995, McKillop et al. 1995), or third, localities could create their own local currencies to enable local people both to provide local work and to buy local goods and services, thus retaining income within the area. To do this, as chapter 8 showed, Local Exchange and Trading Systems (LETS) can be employed. These, to reiterate, are local associations whose members make offers of, and requests for, goods and services and then exchange them priced in a local unit of currency. In theory, these overcome many of the problems with existing policy instruments for local purchasing. Conventional approaches to local purchasing have low multiplier effects because even if a person purchases locally, either the business is often externally owned so some of the money will leave the area, or the income will be used to purchase further inputs which are themselves produced outside the area. In a LETS, however, the local currency can be used only to purchase further goods and services within the system. None of it can leave the locality, nor can it be used to purchase goods and services outside the area since it has no 'value' external to the local association. The effect is that it creates a 'closed system' so that none of the local money leaks out of the area (Davis and Davis 1987, Williams 1996e). In consequence, LETS have been widely advocated as a means by which local purchasing can be encouraged (Dobson 1993, Greco 1994, Lang 1994, Offe and Heinze 1992, Williams 1996 d, e, g). Williams (1996f) has revealed that LETS are exponentially growing. In 1992, there were just 4 LETS in the UK, but by 1995, there were some 350 trading, with around 30000 members. All the signs, moreover, are that they are continuing to develop.

Localizing work to meet local demand

A fourth and final principle for implementing this more locally oriented economic development strategy is to localize work so that it meets local demand. To achieve this, first, a policy of localizing employment could be pursued so as to maximize the number of local people obtaining jobs in the locality, although this is problematic with EU legislation such as in the realm of EU-funded projects and competitive tendering. Such a policy, nevertheless, is a relatively simple tool which can be integrated into a wide range of tenders and contracts by interested parties.

Second, and more widely, a policy of encouraging informal economic activity could be pursued. Although there are many instances in a Third World context where such a policy has been adopted to create work to meet local needs, examples in the advanced economies are notable by their absence. At the time of writing, the only known exception is Hackney Council in London, which is considering harnessing the informal sector on some of its most deprived estates. Such an approach, however, is in its infancy, and few have even begun to think through how the informal sector can be harnessed to meet local needs and wants. Just as consumer services have been consigned to a dependency status, the myth which pervades much academic discourse towards the informal sector is that it is left over from a previous economic era and that economies as they advance shift activity from their informal to

CONCLUSIONS

their formal sector. Neither proposition has yet been seriously investigated. Until such time as they are proven, there is a need to maintain an open mind about harnessing the informal sector as a way of localizing work and promoting local economic development.

To evaluate whether these principles of local ownership, import substitution, local control of money and localizing work to meet local demand promote local revitalization in practice is difficult. Unfortunately, there are few experiences to draw upon. Although the Saint Paul's Homegrown Economy Project (HEP) in Minnesota (Imbroscio 1995, Judd and Ready 1986) attempted to shift from a reliance on imports to greater self-reliance, as Imbroscio (1995: 858) concludes, this 'never matured beyond the status of a symbolic gesture'. The current proposal by 'Forum for the Future', nevertheless, to use West Yorkshire as a crucible for such an approach will provide a much richer source of information over the coming years for those wishing to evaluate this more holistic approach towards local economic development. At the time of writing, however, this remains in its early planning stages.[4] The effectiveness of such localization strategies, therefore, must remain in the realms of theoretical discourse for the immediate future.

CONCLUSIONS

Since the 1980s, a large number of academic commentators have put much effort into elevating the status of producer services and, in so doing, changed the perception of practitioners concerning their role in local economic development. This book has shown that it is now time for a similar drive to occur in the realm of consumer services. There are, at present, many academics working disparately on individual consumer service industries but, as yet, no coordinated endeavour to dispel the myth of consumer services as a dependent sector. It is hoped that this book has provided one impetus to start this process.

By revealing that consumer services are basic activities in the sense that they generate external income, it has brought into focus the inherent weakness of the basic/dependent sector dualism which lies at the heart of economic base theory, the principal conventional conceptual tool used to formulate local economic policy and theorize local economic development. Furthermore, the analysis of the contributions of consumer services to local economic development in both the individual sub-sector and the locality studies has called into question the related notion that there is a hierarchy of sectors based upon their ability to export. The outcome of the revelation that it is not external income generation in itself which causes local economic development but, rather, also leakage prevention and the consequent inter-linkages between the various sectors of the local economy is that a more holistic approach towards local economic development has been advocated based upon the strategies of import substitution, local ownership, local control of money and localizing work to meet local demand. The study of consumer services, therefore, far from being of little use as many service-sector researchers and economic development theoreticians have sometimes assumed, has actually enabled the nature and function of both consumer services and local economic development to be fundamentally reconceptualized. It is to be hoped, therefore, that the turn of the millennium will see more research on this sector, which employs some 40 per cent of all employees, as well as more attention paid to inward-looking strategies which concentrate on creating the synergies between all economic activities which are necessary to meet local people's needs and wants. If this is achieved, then local economic development will surely regain what can and should be its aim: making localities a better place in which to live, work and do business.

NOTES

1 INTRODUCTION

1 The problem is that as services have grown, the number of classificatory sub-categories in this 'residual' category has had to expand rapidly. The UK Standard Industrial Classification (SIC), for example, was first introduced in 1948. After its first revision in 1958, when service employment exceeded that in manufacturing, there were 32 service divisions compared with 100 manufacturing categories. By the 1980 SIC revisions, when service employment was more than double that of manufacturing, 102 service divisions existed compared with 210 in manufacturing (Marshall and Wood 1995), and in the 1992 SIC, when well over four times as many people worked in services as in manufacturing, 243 service divisions had been created compared with 286 manufacturing divisions (Central Statistical Office 1995).

2 However, given that each and every one of these classificatory schemes attempts to group together activities which are defined as services only because of what they are not, all have had problems. Take, for example, the occupation-based classification scheme. Hepworth (1989) classifies service occupations by the extent of information processing and divides services into either information producers (for example chemists, lawyers, accountants); processors (for example office supervisors); distributors (for example teachers, librarians); or infrastructure workers (for example computer operators, printers). However, this definition completely excludes some workers traditionally defined as service workers such as shop assistants and bus drivers.

3 However, it should be noted that since this way of classifying services was developed, the 1980 SIC index has been superseded by the 1992 SIC index. Given that historical data using the 1992 SIC index were not available at the time of writing, and to create continuity with previously published work, this book mostly, although not exclusively, uses the Marshall *et al.* (1988) classificatory scheme based on the 1980 SIC index to categorize services. This makes little difference to overall findings in terms of the size of the consumer services sector or its constituent parts. However, it does mean that some 1993 Census of Employment data have not been used fully. A useful future exercise

will be to use input/output tables to examine the proportion of different services industries fulfilling final and intermediate demand using the 1992 SIC index so as to produce a revised classification of producer, consumer, mixed producer/consumer and public services.

4 It is doubtless the case that if sufficient evidence were available on other consumer service industries such as medical services, school education and welfare services, the nature and extent of their contributions to local economic development would be very similar to those considered in this book. At present, however, there is inadequate information readily available on these other consumer services to permit anything but a superficial analysis of their role in revitalizing local economies.

2 DOES MANUFACTURING MATTER? DEINDUSTRIALIZATION AND TERTIARIZATION IN THE GLOBAL ECONOMY

1 Although later chapters will heavily criticize the economic base theory approach, which defines the 'motor' of local economic development as those sectors which generate external income, it is here used to provide the reader with an overview of the relative importance of manufacturing and services as export-oriented industries since this is a pre-requisite for understanding their fuller contributions to local economic development discussed later.

2 So far as developing nations are concerned, their smaller service sector can in part be explained by the existence of a larger informal sector which fulfils many of the functions undertaken by the formal service sector in advanced economies (Daniels 1993, Gilbert 1994, Rakowski 1994).

3 See chapter 13 for further discussion of the relationship between export-propensity and firm size, sub-sector of services and ownership.

4 For example, respondents may be asked whether or not they have extra-area sales or sources of particular services, or for percentages of extra-area sales or sources. Ashton and Sternel (1978), for example, report the share of service respondents who claimed that 10–50 per cent of their revenues derived from clients outside New England and the share claiming that over 50 per cent were externally sourced. Other studies report the share of extra-regional revenues averaged across all respondents (Michalak and Fairbairn 1993, Harrington and Lombard 1989, Williams 1996a). Sometimes, moreover, averages are weighted by the relative size of the respondents by revenue size (Van Dinteren 1987) or by employment size (Beyers and Alvine 1985).

3 PRODUCER SERVICES AND ECONOMIC REGENERATION

1 In a study of producer services in Montreal, for example, Coffey (1995b) follows conventional wisdom in defining both the finance,

insurance and real estate (FIRE) and business-service sectors as producer services. His own results display, however, that 63.3 per cent of the gross revenue of FIRE services in Montreal derives from households. Indeed, 17.3 per cent of FIRE businesses receive all their revenue and 51.8 per cent obtain 76–100 per cent from final demand. Therefore, some industries are frequently allocated to the producer services sector, despite being essentially consumer services.

2 See Table 4.1 for more detailed information on the growth rates of producer services compared with other economic activities in Britain.

4 CONSUMER SERVICES: A MOTOR FOR LOCAL ECONOMIC DEVELOPMENT?

1 When those sub-sectors of consumer services most rapidly declining are examined, a further tendency in the direction of contemporary society is revealed. The fastest declining are repair of other consumer goods and footwear/leather repairs. This is doubtless a response to the twin processes of technological advance and the ascendency of a 'throwaway' culture and reflects the shift away from, rather than towards, sustainable development in contemporary society (Haughton and Hunter 1994).

2 This, it should be noted, was the case before the advent of Meadowhall regional shopping centre in Sheffield, which is likely to have reinforced this trend further.

5 TOURISM AND ECONOMIC REGENERATION

1 However, such oligopolistic tendencies in the hotel sector are not evenly distributed across space. Although in the US, the ten largest chains account for 40 per cent of all rooms in 1988 (Debbage 1990), in western Europe, TNCs control less than 3 per cent of the total hotel capacity (Urry 1990).

2 The proposed merger at the time of writing between British Airways and American Airlines will further reinforce the oligopolistic structure of the airline industry in both the European and North American markets.

3 One reason why such a large proportion of tourists are European is the contrasting work cultures in advanced economies. In the USA, holiday entitlement restricts tourism since it is only about two weeks. In Japan, similarly, it is only two weeks, of which only eight days is often taken, although this is changing. In the Netherlands and Germany, in contrast, it can be seven to eight weeks (Jefferson 1995).

4 Of course, exports alone are only half the story, as British tourists also go abroad. The deficit in the balance of payments attributable to tourism in 1988 was around £2035 million. However, until the late 1980s, the balance of payments on tourism had only been in deficit for eight years out of the last twenty (1981–4 and 1986–9), and it has been

estimated that for every 1 per cent increase in the market share of world tourism, an additional £1000 million would come into Britain (House of Commons Employment Committee 1990).

5 This emphasis on the positive economic impacts also occurred in both the sports and cultural industries sectors, albeit slightly later than in tourism (see chapters 6 and 9 for explanations of why this tendency occurred in these sectors).

6 So too has the tourist industry itself responded to this increasing interest in sustainable development. The Earth Centre in the Dearne Valley near Doncaster, for example, is a £114 million project located in an ex-mining area which intends to promote the notion of sustainability (Smales, Prior and Burley 1996). According to the economic impact analysis, this is forecast to receive 2.5 million visitors by the year 2000 and, if it does it will be one of the most popular tourist attractions in Britain (Williams *et al.* 1993).

7 Private companies will not engage in place promotion because it is likely that their competitors will reap the same rewards as themselves from such costs.

6 SPORT AND THE REJUVENATION OF LOCAL ECONOMIES

1 The value-added part of economic activity is the difference between the value of the goods and services produced and the costs of the material inputs and services used in their production and distribution.

2 See chapter 11, moreover, for an assessment of the economic impact of the World Student Games on the Sheffield and regional economy.

3 See chapters 10 and 11 for attempts to assess the short- and long-term impacts of the 1990 Glasgow City of Culture and the Sheffield World Student Games 'hallmark' events respectively.

4 Similarly, in other European nations such as Italy, the municipality owns the ground (for example the San Siro in Milan, which provides a home from AC Milan and Inter, whilst Turin's Stadio Nuovo Comunale is used by both Juventus and Torino).

5 Indianapolis's efforts in the realm of sports are similar in nature to Louisville's emphasis on the arts to anchor downtown development (see Whitt 1988).

7 UNIVERSITIES AND LOCAL ECONOMIC DEVELOPMENT

1 Because of the problems in identifying university employment in the 1993 Census of Employment data, the broader category of higher education is examined here. Not all higher education employment, however, is in the university sector: 50 Colleges of Higher Education were funded by the Higher Education Funding Council for England (HEFCE) in 1993/4 and 288 institutions, mainly further education

colleges, taught 42006 higher education students through franchising arrangements with 74 higher education institutions (HEFCE 1995).

2 A Local Labour Market Area (LLMA) is defined as an area in which the majority of people who live within them work and shop (Coombes *et al.* 1982).

3 In part, this may be because universities have not 'down-sized' to the same extent as other large firms, for example by pursuing out-sourcing policies. Instead, universities have continued to retain many functions in-house and have increased the numbers of both employees and students. Although some sub-contracting in the university sector in the form of 'bridging courses' and franchising teaching to further education colleges is increasing, it is relatively minor compared with the overall increase in 'in-house' employment and student numbers.

4 For an excellent critical appraisal of the use of employment multipliers in measuring the local economic impacts of universities, see Lincoln *et al.* (1994).

5 Implicit in this definition of technology transfer is that an invention is passed on. In reality, scientific and technological knowledge flows two ways between universities and industry, much in a non-commercial form, through learning by doing or in the form of individuals. It is not a one-way relationship.

6 Universities are also vacation conference venues, with 6.4 per cent of the conference market in terms of days, which generated £75 million in 1990/1. Thirty universities also run year-round management training centres (15 per cent of the UK's independent management training centres), but have yet to achieve a significant stake in the lucrative and less price-sensitive corporate sector because they lack top-standard amenities, especially catering (Goddard *et al.* 1994).

7 During the past twelve months, for example, the author has been approached by two local authorities, the police service, a private-sector company and three not-for-profit agencies seeking either knowledge generation or transfer, non of which wished to pay for the service.

8 RETAILING AND ECONOMIC DEVELOPMENT

1 One notable exception is the edited volume by Wrigley and Lowe (1996).

2 For in-depth analyses of retail growth and change, see Guy (1994) and the edited volume by Wrigley and Lowe (1996).

3 As Hughes (1996) highlights, such concentration is lower in the US, where the top five US chains only have 20 per cent of the total US grocery market. This is due to the strict enforcement of anti-trust legislation.

4 An exception is North-East Lincolnshire Council, whose 'buy local' campaign officer is starting to focus upon the retail sector.

9 CULTURAL INDUSTRIES AND LOCAL REVITALIZATION

1 Several problems result from this definition of the cultural industries. Is education, for example, a cultural industry since it conveys meaning and involves a creative input? Similarly, are producer services such as advertising and architecture also cultural industries for the same reasons? Indeed, given the strong symbolic content of most products in contemporary economies, can all industries be viewed as cultural industries? Here, nevertheless, the conventional working definition of the cultural industries as comprising the pre-electronic performing and visual arts and the electronic cultural industries (for example film, photography, music, publishing, design and fashion) is adopted.

2 This is similar to estimates in other countries. Sadek (1995), for example, estimates that 1.9 per cent of the employed workforce in Ireland are employed in cultural industries. However, Urban Cultures Ltd (1994) argue that SIC data seriously under-represent the number of people working in the cultural industries since many are self-employed or even have a day-job in a different industry.

3 The employment multiplier effects of the cultural industries, at 0.33–0.5, are much lower than those identified for universities in chapter 7 of 1.1–1.8. This is mostly due to the lower level of leakage in universities compared with cultural industries. In the former, both labour inputs and other goods and services are more likely to be local than in the latter, where not only is labour more likely to be external (for example in the performing arts) but so too are the other good and service inputs (for example in the printing and publishing industry).

4 It should be noted, however, that although such festivals are pitched at national and international audiences, many were originally aimed at local citizens.

10 GLASGOW: A CITY OF CULTURE?

1 See Part I of Pacione (1995) for an in-depth chronological treatment of the main agents, processes and patterns underlying the development of Glasgow from its pre-urban origins until the close of the nineteenth century and Part II for an analysis of the socio-spatial development of Glasgow in the twentieth century.

2 By 1994, overseas companies employed around a quarter of all manufacturing employees in Scotland (Foreign and Commonwealth Office 1996).

3 See chapter 6 of Pacione (1995) for an in-depth discussion of Glasgow's changing urban economy.

4 This is true not only in terms of jobs but also in terms of the type of job, which did not match the needs or skills of the redundant labour force.

5 See chapter 9 of Pacione (1995) for a more in-depth examination of these socio-economic problems.

6 Subsequent awards have gone to Dublin (1991), Madrid (1992), Nuremberg (1993) and Lisbon (1994).
7 For details of Glasgow's image, see chapter 10 of Pacione (1995).

11 BEYOND STEEL CITY: CONSUMER SERVICES AND ECONOMIC DEVELOPMENT IN SHEFFIELD

1 Contrary to popular opinion, however, Sheffield now makes more steel than ever before. In 1942, Sheffield made a record 1.9 million tonnes of steel in 176 furnaces. In 1992, just seven furnaces produced 2.25 million tonnes, but with only an eighth of the 1942 labour force (Hamilton-Fazey 1993a, Wright 1996).
2 As Taylor et al. (1996) highlight, although this was later replaced by Sheffield City Liaison group, both these partnership bodies have acted less as a 'growth coalition' in the Molotch and Logan (1985) sense and more as a 'rescue squad' to minimize the economic and social problems resulting from deindustrialization.
3 Interest in the World Student Games seems to have increased since Sheffield hosted them. Three firm bids were received for the 1993 Games, awarded to Buffalo in the USA and Fukoka in Japan paid FISU $3 million for the rights to stage the 1995 games (Foley 1991).
4 The result was that in 1995, retail sales totalled £450 million in the city centre's 500 outlets (compared with £325 million at Meadowhall's 284 outlets), a 5 per cent increase in turnover on 1994 (Lambert 1996).
5 Nevertheless, it is not so one-sided as this suggests. The recent £6 million pilot project, funded by the EU, for a regional technopole based on traditional manufacturing industries is but one example of how local economic development continues to pay attention to manufacturing.

12 CORPORATE CITY: THE ROLE OF CONSUMER SERVICES IN THE REJUVENATION OF LEEDS

1 These have been analysed rather than other sectors because they are the principal consumer services upon which local economic policymakers have focused their attention. Other sub-sectors of consumer services such as sport, higher education and medical services doubtless have similar economic impacts on the Leeds economy.
2 In Sheffield with its production-oriented approach, meanwhile, the changes in employment in the various sub-sectors of the cultural industries are nearly a mirror-image, with rapid growth in production-oriented industries and much slower growth or decline in its consumption-oriented cultural industries.
3 One reason for this may well be that the retail sector, similar to other industrial sectors, benefits from economies of scale with regard to both individual retail units in particular, and shopping centres more generally, which means that as retail floorspace increases, retail job numbers do not witness a parallel linear rise.

253

13 CONSUMER SERVICES AND RURAL REGENERATION: A CASE STUDY OF THE FENS

1 Between 1960 and 1987 in Britain, for example, manufacturing jobs increased by 19.7 per cent in rural areas, compared with an overall national decline of 37.5 per cent (DoE and MAFF 1995).

2 However, relatively few rural authorities compared with urban areas have either an arts development officer or strategy which results in a vicious circle of fewer activities and thus less chance of a strategy being formulated and a person being given responsibility for such activities (Williams *et al.* 1995).

3 The extent to which this is occurring should not be over-emphasized. Overall, the impression portrayed in the rural services literature is that services are under-represented in rural areas and that they have less of a role to play in the rural landscape than in urban areas.

4 This section is an abridged and modified version of the empirical evidence reported in Williams (1996a).

5 Examining the register of commercially rated premises reveals that larger establishments had higher response rates than smaller establishments and that service-sector businesses (40 per cent response rate) were slightly more likely than manufacturing (33.3 per cent) and construction (35.8 per cent) to respond. In order to allocate every business establishment to a four-digit activity code using the 1980 SIC index, each establishment was requested to describe the nature of the business in which it was engaged. Marshall's (1988) classification scheme outlined in chapter 1 was then used to allocate each establishment to one of primary, manufacturing, producer services, consumer services or mixed producer/consumer services. To discover the geographical structure of the markets of Fenland businesses, each establishment was requested to provide a breakdown of the percentage of their suppliers and customers who were local, from the rest of East Anglia, from the rest of the UK (excluding East Anglia) and from the rest of the world. In view of the perceived sensitivity of such data to the businesses involved, proportions rather than absolute data on customers and suppliers were requested. All survey respondents provided these data, and nearly all stated within a margin of ten percentage points the source of customers and/or suppliers. Indeed, most responses were provided to within five percentage points and many were within one percentage point. From this, it can be assumed that Fenland businesses have a keen awareness of the geographical source of their suppliers and consumers and that the information provided is therefore relatively reliable.

14 LONDON: THE ROLE AND CHARACTER OF CONSUMER SERVICES IN A GLOBAL CITY

1 The other cities with high concentrations of the top 500 companies headquarters are Tokyo (with 34), Paris (26), Chicago (18), Essen (18),

Osaka (15), Los Angeles (14), Houston (11), Hamburg (10) and Dallas (10).

2 However, only two of these are the international headquarters of the company in question. Other international headquarters are in Japan, the USA or Germany, reflecting the origins of the companies concerned (Urban Cultures Ltd 1994).

3 Clark and McNicholas (1996), moreover, identify a similar professionalization in Los Angeles, with continuous growth at the top end both absolutely and relatively and a shrinking of the less-skilled sector.

4 Indeed, given that the average wage in Britain for personal and protective services is £296.10 for men and £198.70 for women, whilst plant and machine operatives receive £293.70 (men) and £201.50 (women) (New Earnings Survey 1995: Tables 122 and 123), serious questions need to be raised about the assumption that the consumer services sector is necessarily a lower-wage sector than manufacturing.

15 CONCLUSIONS: RETHINKING CONSUMER SERVICES AND LOCAL ECONOMIC DEVELOPMENT

1 Indeed, the mere fact that certain activities and businesses within a particular sector do not earn external income is no reason for rejecting it as a basic-sector. As witnessed in the Fens, the basic-sector orientation of the manufacturing sector is far from universal, with 11.9 per cent of manufacturing firms in this local economy earning none of their income from outside the locality and only 63.1 per cent having over half their customers outside the locality (see chapter 13).

2 This process of concentrated decentralization is not solely confined to producer and consumer service industries. Marshall (1996) identifies a similar tendency in central government public services whereby there has been a concentrated decentralization around London and the more recent reorganization has led to the separation of service provision from policy-making in a periphery/core pattern.

3 It is important here to distinguish between self-reliance and self-sufficiency. Self-sufficiency means using only local products and services, whereas self-reliance means using local goods and services whenever possible and importing only when these products and services cannot be made available locally.

4 For more up-to-date information on this proposal by 'Forum for the Future', contact either the author or Forum for the Future, Thornbury House, 18 High Street, Cheltenham GL50 1DZ.

REFERENCES

Aldskogius, H. (1995) 'Ice hockey and "place": a great club in a small town', in J. Bale (ed.) *Community, landscape and identity: horizons in a geography of sports*, Keele: Department of Geography Occasional Paper no. 20, University of Keele.

Allen, J. (1988) 'Towards a post-industrial economy?', in J. Allen and D. Massey (eds) *The economy in question*, London: Sage.

Archer, B. (1977) *Tourism multipliers: the state of the art*, Bangor: University of Wales Press.

—— (1982) 'The value of multipliers and their policy implications', *Tourism Management* 3, 4: 236–41.

—— (1984) 'Economic impact: misleading multiplier', *Annals of Tourism Research* 11, 3: 517–18.

Arlidge, J. (1996) 'Glasgow's fix of culture and couture', *The Observer* 19 May: 14.

Armstrong, H. W. (1993) 'The local income and employment impact of Lancaster University', *Urban Studies* 30: 1653–68.

Armstrong, H. W., Darrall, J. and Grove-White, R. (1994) *Building Lancaster's future: economic and environmental implications of Lancaster University's expansion to 2001*, Lancaster: Department of Economics and Centre for the Study of Environmental Change, Lancaster University.

Arnold, A. (1986) 'The impact of the Grand Prix on the transport sector', in J. P. A. Burns, J. H. Hatch and T. J. Mules (eds) *The Adelaide Grand Prix: the impact of a special event*, Adelaide: Centre for South Australian Economic Studies.

Ashton, D. J. and Sternel, B. K. (1978) *Business services and New England's export base*, Boston: Research Department, Federal Reserve Bank of Boston.

Ashworth, G. J. and Ennen, E. (1995) 'Culture to the rescue: art as an instrument in urban economic policy', Paper presented to Association of European Schools of Planning Conference, Glasgow, August.

Aylen, J. (1984) 'Prospects for steel', *Lloyds Bank Review* 152: 13–19.

Baade, R. A. and Dye, R. F. (1988a) 'Sports stadiums and area development: a critical review', *Economic Development Quarterly* 2, 3: 265–75.

256

REFERENCES

—— (1988b) 'An analysis of the economic rationale for public subsidisation of sports stadiums', *Annals of Regional Science* 22, 2: 37–47.

—— (1990) 'The impact of stadiums and professional sports on metropolitan area development', *Growth and Change* 21, 2: 1–14.

Bachtler, J. and Davies, P. L. (1989) 'Economic restructuring and services policy', in D. Gibbs (ed.) *Government policy and industrial change*, London: Routledge.

Bailly, A. S. (1995) 'Producer services research in Europe', *The Professional Geographer* 47, 1: 70–3.

Baim, D. (1990) *Sports stadiums as wise investments: an evaluation*, Detroit, MI: Heartland Institute Policy Study no. 32.

—— (1992) *The sports stadium as a municipal investment*, Westport, CT: Greenwood Press.

Bale, J. (1991) 'Sport and local economic development: a review of the literature', Paper presented to the Annual Conference of the Institute of British Geographers, University of Sheffield, January.

—— (1993) *Sport, space and the city*, London: Routledge.

Bale, J. and Moen, O. (1995) (eds) *The stadium and the city*, Keele: Keele University Press.

Barford, J. (1987) 'Metro Centre, Gateshead', *Housing and Planning Review* 43, 3: 10–11.

Bassett, K. (1993) 'Urban cultural strategies and urban regeneration: a case study and critique', *Environment and Planning A* 25: 1773–88.

Bedwell, P. (1993) 'The effectiveness of town centre management', unpublished Diploma in Town and Regional Planning dissertation, Leeds: Leeds Metropolitan University.

Bennington, J. (1986) 'Local economic strategies: paradigms for a planned economy', *Local Economy* 1: 13–21.

Berg, L. van den, van der Borg, J. and van der Meer, J. (1995) *Urban tourism: performance and strategies in eight European cities*, Aldershot: Avebury.

Beyers, W. B. (1993) 'Producer services', *Progress in Human Geography* 16, 4: 578–83.

Beyers, W. B. and Alvine, M. J. (1985) 'Export services in postindustrial society', *Papers of the Regional Science Association* 57: 33–45.

Bianchini, F. (1989) 'Cultural policy and urban social movements: the response of the new left in Rome (1976–1985) and London (1981–1986)', in P. Bramham, I. Henry, H. Mommaas and H. van der Poel (eds) *Leisure and Urban Processes*, London: Routledge.

—— (1993a) 'Remaking European cities: the role of cultural policies', in F. Bianchini and M. Parkinson (eds) *Cultural policy and urban regeneration: the West European experience*, Manchester: Manchester University Press.

—— (1993b) 'Culture, conflict and cities: issues and prospects for the 1990s', in F. Bianchini and M. Parkinson (eds) *Cultural policy and urban regeneration: the West European experience*, Manchester: Manchester University Press.

—— (1995) 'Night cultures, night economies', *Planning Practice and Research* 10, 2: 121–6.

257

REFERENCES

Bianchini, F., Montgomery, J., Fisher, M. and Warpole, K. (1988) *City centres, city cultures*, Manchester: Centre for Local Economic Strategies.

Blake, C., McDowell, S. and Devlin, J. (1979) *The 1978 Open Golf championship at St Andrews: an economic impact study*, Edinburgh: Scottish Academic Press.

Blau, J. (1989) *The shape of culture*, Cambridge: Cambridge University Press.

Bleaney, M. F., Binks, M. R., Greenway, D., Reed, G. V. and Whynes, D. K. (1992) 'What does a university add to its local economy?', *Applied Economics* 24: 304–11.

Bloomfield, J. (1993) 'Bologna: culture, citizenship and quality of life', in F. Bianchini and M. Parkinson (eds) *Cultural policy and urban regeneration: the West European experience*, Manchester: Manchester University Press.

Bloomquist, K. (1988) 'A comparison of alternative methods for generating economic base multipliers', *Regional Science Perspectives* 8: 58–99.

Bluestone, H. and Hession, J. (1986) *Patterns of change in the metro and nonmetro labor force since 1979*, Washington, DC: US Department of Agriculture Economic Research Service.

Bluestone, S. and Harrison, B. (1982) *The deindustrialization of America*, New York: Basic Books.

Blumenthal, D. (1994) 'Growing pains for new academic/industry relationships', *Health Affairs* 13, 3: 176–93.

Booth, P. and Boyle, R. (1993) 'See Glasgow, see culture', in F. Bianchini and M. Parkinson (eds) *Cultural policy and urban regeneration: the West European experience*, Manchester: Manchester University Press.

Bourdieu, P. (1985) *Distinction: social critique of the judgement of taste*, London: Routledge and Kegan Paul.

Boyer, M. C. (1992) 'Cities for sale: merchandising history at South Street Seaport', in M. Sorkin (ed.) *Variations on a theme park: a new American city and the end of public space*, New York: Hill and Wang.

Boyle, M. and Hughes, G. (1991) 'The politics of the representation of the "real": discourses from the Left on Glasgow's role as European City of Culture 1990', *Area* 23, 3: 217–28.

—— (1994) 'The politics of urban entrepreneurialism in Glasgow', *Geoforum* 25, 4: 453–70.

Bramwell, B. (1993) 'Planning for tourism in an industrial city', *Town and Country Planning* 62, 1/2: 17–19.

—— (1995) *An event-led city and tourism marketing strategy for Sheffield*, Sheffield: Destination Sheffield.

Bramwell, B. and Lane, B. (1994) (eds) *Rural tourism and sustainable rural development*, Clevedon: Channel View Publications.

Bramwell, B. and Rawding, L. (1994) 'Tourism marketing organizations in industrial cities', *Tourism Management* 15, 6: 425–34.

—— (1996) 'Tourism marketing images of industrial cities', *Annals of Tourism Research* 23, 1: 201–21.

Brewers and Licensed Retailers Association (1994) *A real alternative: memorandum to HM Treasury*, London: Brewers and Licensed Retailers Association.

—— (1995) *A smugglers' charter: the cross-channel beer trade*, London: Brewers and Licensed Retailers Association.

British American Arts Association (1989) *Arts and the changing city: an agenda for urban regeneration*, London: British American Arts Association.

Britton, S. (1978) 'International tourism and indigenous development objectives: a study with reference to the West Indies', unpublished Ph.D. thesis, University of Minnesota, Minneapolis, MN.

—— (1991) 'Tourism, capital and place: towards a critical geography of tourism', *Environment and Planning D* 9: 451–78.

Bromley, R. D. F. and Thomas, C. (1993) 'The retail revolution, the carless shopper and disadvantage', *Transactions* 18, 2: 222–36.

Burgan, B. and Mules, T. (1992) 'Economic impact of sporting events', *Annals of Tourism Research* 19: 700–10.

Burns, J. P. A. and Mules, T. J. (1986) 'An economic evaluation of the Adelaide Grand Prix', in G. J. Syme, B. J. Shaw, P. M. Fenton and W. S. Mueller (eds) *The planning and evaluation of hallmark events*, Aldershot: Avebury.

Burns, J. P. A., Hatch, J. H. and Mules, T. J. (1986) *The impact of a special event*, Adelaide: Centre for South Australian Economic Studies, University of Adelaide.

Butler, R. (1993) 'Pre- and post-impact assessment of tourist development', in D. Pearce and R. Butler (eds) *Tourism research: critiques and challenges*, London: Routledge.

Cambridge City Council (1993) *Tourism 2000: action plan for Cambridge*, Cambridge: Cambridge City Council.

Campbell, M. (1996) 'The Leeds economy: trends, prospects and challenges', in G. Haughton and C. C. Williams (eds) *Corporate city? Partnership, participation and partition in urban development in Leeds*, Aldershot: Avebury.

Cappelin, R. (1989) 'The diffusion of production services in the urban system', *Revue d'Economie Regionale et Urbaine* 4: 625–40.

Castells, M. and Hall, P. (1994) *Technopoles of the world*, London: Routledge.

Castro, C., McMullan, W. E. and Vesper, K. H. (1987) The venture generating potential of a university', *Journal of Small Business and Entrepreneurship* 5, 2: 31–40.

Cater, E. (1995) 'Environmental contradictions in sustainable tourism', *Geographical Journal* 161, 1: 21–8.

Central Statistical Office (1993) *Census of Employment 1991*, London: CSO.

—— (1995) *Census of Employment 1993*, London: CSO.

Centre for Advanced Studies in the Social Sciences (1995) *The economic impact of sport in Wales: a report prepared for the Sports Council of Wales*, Cardiff: University of Wales College Cardiff.

Centre for Applied Business Research (1987) *The impact of the America's Cup defence series, 1986–87*, Perth: Centre for Applied Business Research.

Chesterton (1993) 'The MetroCentre', mimeo.

Chrisman, J. J., Hynes, T. and Fraser, S. (1995) 'Faculty entrepreneurship and economic development: the case of the University of Calgary', *Journal of Business Venturing* 10, 4: 267–81.

Christaller, W. (1963) 'Some considerations of tourism location in Europe', *Papers and Proceedings of the Regional Science Association* 12:95–105.

—— (1966) *Central places in southern Germany*, Englewood Cliffs, NJ: Prentice-Hall.

Churchman, C. (1995) 'Sports stadia and the landscape: a review of the impacts and opportunities arising as a result of the current development of football grounds', *Built Environment* 21, 1: 6–24.

City of Chicago Department of Economic Development (1986) *The economic impact of a major league baseball team on the local economy*, Chicago: City of Chicago Department of Economic Development.

Clapp, D. (1987) 'Going for gold', *Initiatives* June: 8–9.

Clark, C. A. (1940) *The conditions of economic progress*, London: Macmillan.

Clark, W. A. V. and McNicholas, M. (1996) 'Re-examining economic and social polarisation in a multi-ethnic metropolitan area: the case of Los Angeles', *Area* 28, 1: 56–63.

Clout, H. (1993) *European experience of rural development*, London: Rural Development Commission.

Coccossi, H. and Nijkamp, P. (1995) (eds) *Sustainable tourism development*, Aldershot: Avebury.

Coe, N. M. (1996) 'Uneven development in the UK computer services industry since 1981', *Area* 28, 1: 64–77.

Coffey, W. J. (1995a) 'Producer services research in Canada', *The Professional Geographer* 47, 1: 74–81.

—— (1995b) 'Forward and backward linkages of producer service establishments: evidence from the Montreal metropolitan area', Paper presented to AAG Annual Meeting session of *Producer services in an age of flexibility*, Chicago, March.

Coffey, W. J. and Polese, M. (1987) 'Trade and location of producer services: a Canadian perspective', *Environment and Planning A* 19: 597–611.

Cohen, R. (1981) 'The new international division of labour: multinational companies and urban hierarchy', in A. J. Scott and M. Dear (eds) *Urbanization and planning in capitalist society*, London: Methuen.

Comedia (1991) *Out of hours: a study of economic, social and cultural life in twelve town centres in the UK*, Stroud: Comedia.

—— (1992) *The position of culture in London*, Stroud: Comedia.

Commission of the European Communities (1990) *Green paper on the Urban Environment*, Brussels: COM (90) 218.

Committee of Vice-Chancellors and Principals (1995) *The economic impact of international students in UK higher education*, London: Committee of Vice-Chancellors and Principals.

Cooke, P. (1990) *Back to the future: modernity, post-modernity and locality*, London: Unwin Hyman.

Coombes, M. G., Dixon, J. S., Goddard, J. B., Openshaw, S. and Taylor, P. J. (1982) 'Functional regions for the Population Census of Britain', in D. T. Herbert and R. J. Johnston (eds) *Geography and the urban environment* 5, London: Wiley.

Corliss, R. (1992) 'Build it and they might come', *Time* 1: 50–2.

Cox, K. (1979) *Location and public policy*, Oxford: Blackwell.

Creative Consumer Co-operative Ltd (1995) *Out of this World: the opportunity*, Newcastle upon Tyne: Creative Consumer Co-operative Ltd.

REFERENCES

Crompton, J. L. (1995) 'Economic impact analysis of sports facilities and events: eleven sources of misapplication', *Journal of Sports Management* 9: 14–35.

Cwi, D. (1982) 'The focus and impact of arts impact studies', in C. Vilette and R. Taqqu (eds) *Issues in supporting the arts*, Ithaca, NY: Cornell University, Graduate School of Business and Public Administration.

Dabinett, G. (1991) 'Local policies towards industrial change: the case of Sheffield's Lower Don Valley', *Planning Practice and Research* 6, 1: 13–18.

Dalton, I. G. (1995) 'The future for science parks', in H. Armstrong, T. Brice and B. Jackson (eds) *Cities of learning: papers from a Conference held on 20 & 21 April 1995*, London: Committee of Vice-Chancellors and Principals.

Damer, S. (1990) *Glasgow: going for a song*, Lawrence and Wishart, London.

Daniels, P. W. (1985) *Service industries: a geographical appraisal*, Andover: Methuen.

—— (1986) 'Foreign banks and metropolitan development: a comparison of London and New York', *Tijdschrift voor Economische en Sociale Geografie* 77: 269–87.

—— (1991) 'Some perspectives on the geography of services', *Progress in Human Geography* 15: 37–46.

—— (1993) *Service industries in the world economy*, Oxford: Blackwell.

—— (1995a) 'The locational geography of advanced producer services firms in the United Kingdom', *Progress in Planning* 43, 2–3: 123–38.

—— (1995b) 'Producer services research in the United Kingdom', *The Professional Geographer* 47, 1: 82–6.

Dauncey, G. (1988) *After the crash: the emergence of a rainbow economy*, London: Green Print.

Davis, H. C. and Davis, L. E. (1987) 'The Local Exchange Trading System: community wealth creation within the informal economy', *Plan Canada* 20: 238–45.

Debbage, K. (1990) 'Oligopoly and the resort cycle in the Bahamas', *Annals of Tourism Research* 17: 513–27.

Department of Environment (1993) *Town centres and retail development*, Planning Policy Guidance note 6, London: Department of Environment.

—— (1994) *Vital and viable town centres: meeting the challenge*, London: Department of Environment.

—— (1995) *Town centres and retail development*, Planning Policy Guidance note 6, London: Department of Environment.

Department of Environment and Ministry of Agriculture, Fisheries and Food (1995) *Rural England: a nation committed to a living countryside*, London: HMSO.

Department of Sport, Recreation and Tourism (1986) *Economic impact of the World Cup of Athletics held in Canberra in October 1985*, Canberra: Research and Development Section, Department of Sport, Recreation and Tourism.

Derrick, E. and McRory, J. (1973) *Sunderland's self-image after the Cup*, Working paper no. 8, Birmingham: Centre for Urban and Regional Studies, University of Birmingham.

REFERENCES

Destination Sheffield Visitor and Conference Bureau (1994) *1993 Review and Business Plan 1994–5*, Sheffield: Destination Sheffield Visitor and Conference Bureau.

—— (1996) *1995 Review and Business Plan 1996–7*, Sheffield: Destination Sheffield Visitor and Conference Bureau.

Dicken, P. (1992) *Global shift: the internationalization of economic activity*, London: Paul Chapman.

Dill, D. D. (1995) 'University–industry entrepreneurship: the organisation and management of American university technology transfer units', *Higher Education Quarterly* 49, 4: 369–84.

diMaggio, P. (1982) 'Cultural entrepreneurship in 19th century Boston: the creation of an organisational base for high culture in America', *Media, Culture and Society* 4: 35–50.

Dobson, R. V. G. (1993) *Bringing the economy home from the market*, London: Black Rose Books.

Donnison, D. and Middleton, A. (1987) *Regenerating the inner city: Glasgow's experience*, London: Routledge.

Doutriaux, J. (1991) 'University culture, spin-off strategy, and success of academic entrepreneurs at Canadian universities', in N. C. Churchill, W. D. Bygrave, J. G. Covin, D. C. Sexton, D. P. Slevin, K. H. Vesper and W. E. Wetzel (eds) *Frontiers of entrepreneurship research*, Wellesley, MA: Babson College.

Dreeszen, C. (1994) 'Reimagining communities: community arts and cultural planning in America', unpublished dissertation, University of Massachusetts, Department of Landscape Architecture and Regional Planning.

Drennan, M. P. (1992) 'Gateway cities: the metropolitan source of US producer service exports', *Urban Studies* 29: 217–35.

Duffield, B. S. and Vaughan, D. R. (1981) *The economy of rural communities in the National Parks of England and Wales*, Edinburgh: Tourism and Recreation Research Unit, University of Edinburgh.

Eckstein, J., Selwood, S. and Dunlop, R. (1996) *The economics of the cultural sector*, London: Policy Studies Institute.

Economics Research Associates (1984) *Community economic impact of the Olympic Games in Los Angeles*, Los Angeles: ERA.

Ekins, P. and Max-Neef, M. (1992) (eds) *Real-life economics: understanding wealth-creation*, London: Routledge.

Elfring, T. (1989) 'The main features and underlying causes of the shift to services', *Service Industries Journal* 9: 337–46.

English Tourist Board (1991) *Investment in tourism*, London: English Tourist Board.

—— (1993) *Sightseeing in 1992*, London: English Tourist Board.

Erickson, R. A. (1989) 'Export performance and state industrial growth', *Economic Geography* 65: 281–92.

Esparza, A. and Krmenec, A. J. (1994) 'Producer services trade in city systems: evidence from Chicago', *Urban Studies* 31, 1: 29–46.

Euchner, C. C. (1993) *Playing the field: why sports teams move and cities fight to keep them*, Baltimore: Johns Hopkins University Press.

REFERENCES

Farness, D. H. (1989) 'Detecting the economic base: new challenges', *International Regional Science Review* 12: 319–28.

Farrao, J. and Domingues, A. (1995) 'Portugal: the territorial foundations of a vulnerable tertiarisation process', *Progress in Planning* 43, 2–3: 241–60.

Farrar, J. and Smith G. (1995) 'Centre management bandwagon rolls on', *Planning* 1139: 20–1.

Feagin, J. R. and Smith, M. P. (1987) 'Cities and the new international division of labour: an overview', in M. P. Smith and J. R. Feagin (eds) *The capitalist city*, Oxford: Blackwell.

Fieldhouse, N. (1996) '£500 million investment in change for the better', *Sheffield Star: city centre business plan* 14 May: 2.

Fimrie, R. (1992) 'Oh give me a home . . .', *Sports Illustrated* 76, 21: 50–2.

Fisher, A. G. (1935) *The clash of progress and security*, London: Macmillan.

Fletcher, J. (1989) 'Input–output analysis and tourism impact studies', *Annals of Tourism Research* 16, 4: 514–29.

Fletcher, J. and Archer, B. (1991) 'The development and application of multiplier analysis', in C. Cooper (ed.) *Progress in tourism, recreation and hospitality management*, London: Belhaven.

Fletcher, J. and Snee, H. (1989) 'Tourism multiplier effects', in S. Witt and L. Moutinho (eds) *Tourism marketing and management handbook*, Hemel Hempstead: Prentice-Hall.

Florax, R. (1992) *The university: a regional booster? Economic impacts of academic knowledge infrastructure*, Aldershot: Avebury.

Foley, P. (1991) 'The impact of the world student games on Sheffield', *Environment and Planning C* 9, 1: 65–78.

Foley, P. and Masser, F. I. (1988) 'Forecasting and the development of objectives in local economic planning', *The Planner* 74, 5: 21–5.

Foley, P., Hutchinson, J., Harbane, B. and Tait, G. (1996) 'Impact of Toyota on Derbyshire's local economy and labour market', *Tijdschrift voor Economische en Sociale Geografie* 87, 1: 19–31.

Foreign and Commonwealth Office (1995) *Talking points on Britain's economy*, London: Foreign and Commonwealth Office.

—— (1996) *Talking points on Britain's economy*, London: Foreign and Commonwealth Office.

Foster, J. and Woolfson, C. (1986) *The politics of the UCS Work-in*, Lawrence and Wishart, London.

Freitag, T. G. (1994) 'Enclave tourism development: for whom the benefits roll?', *Annals of Tourism Research* 21, 3: 538–54.

Friedmann, J. (1986) 'The world city hypothesis', *Development and Change* 17: 69–83.

Friedmann, J. and Woolf, G. (1982) 'World city formation: an agenda for research and action', *International Journal of Urban and Regional Research* 6: 309–43.

Frost, M. and Spence, N. (1993) 'Global city characteristics and central London's employment', *Urban Studies* 30, 3: 547–58.

Frost-Kumpf, H. A. and Dreeszen, C. (1995) 'Planning for culture? Some spatial aspects of the new field of cultural planning', Paper presented to the Association of American Geographers Annual Meeting, Chicago, March.

Garnham, N. (1990) *Capitalism and communication*, London: Sage.

Gateshead MBC (1990) *MetroCentre Information Note March 1990*, Gateshead: Gateshead Metropolitan Borough Council.

—— (1992) *Retail change on Tyneside and the impact of MetroCentre, Report of the Director of Planning*, Gateshead: Gateshead Metropolitan Borough Council.

Gentle, C. (1993) *The financial services industry: the impact of corporate organisation on regional economic development*, Aldershot: Avebury.

Gershuny, J. (1978) *After industrial society: the emerging self-service economy*, London: Macmillan.

—— (1992) 'Are we running out of time?', *Futures* (January/February): 3–22.

Getz, D. (1991) *Festivals, special events and tourism*, New York: Van Nostrand Reinhold.

Giaratani, F. and McNelis, P. (1980) 'Time series evidence bearing on crude theories of regional growth', *Land Economics* 6: 238–48.

Gibbs, D. (1989) *Government Policy and Industrial Change*, London: Routledge.

Gilbert, A. (1994) Third World cities: poverty, employment, gender roles and the environment during a time of restructuring', *Urban Studies* 31, 4–5: 605–33.

Gillis, W. (1987) 'Can service-producing industries provide a catalyst for regional economic growth?', *Economic Development Quarterly* 1: 249–56.

Gilmer, R. W., Keil, S. R. and Mack, R. S. (1987) 'Export potential of services in the Tennessee valley', *Regional Science Perspectives* 17, 2: 27–41.

Glasgow District Council (1987) *European City of Culture 1990, City of Glasgow District Council, submission for the United Kingdom nomination*, Glasgow: Glasgow District Council.

Glasmeier, A. and Borchard, G. (1989) 'From branch plants to back offices: prospects for rural services growth', *Environment and Planning A* 21: 1565–83.

Glasmeier, A. and Howland, M. (1994a) 'Service-led rural development: definitions, theories and empirical evidence', *International Regional Science Review* 16, 1–2: 197–229.

—— (1994b) *From combines to computers: rural services and development in the age of information technology*, New York: State University of New York Press.

Glennie, P. D. and Thrift, N. (1992) 'Modernity, urbanism and modern consumption', *Environment and Planning A* 10: 423–43.

Glickman, N. J. (1977) *Econometric analysis of regional systems: explorations in model building and policy analysis*, London: Academic Press.

Goddard, J., Charles, D., Pike, A., Potts, G. and Bradley, D. (1994) *Universities and communities*, London: Committee of Vice-Chancellors and Principals.

Goe, W. R. (1994) 'The producer services sector and development within the deindustrializing urban community', *Social Forces* 72, 4: 971–1009.

Goldsmiths' College (1995) *Goldsmiths' College and the local community: the college's impact on Deptford and on South East London*, London: Goldsmiths' College, University of London.

Goss, J. (1993) 'The "magic of the mall": an analysis of form, function and meaning in the contemporary retail built environment', *Annals of the Association of American Geographers* 83, 1: 18–47.

Greco, T. H. (1994) *New money for healthy communities*, Tucson, AZ: Thomas H. Greco.

Green, I. (1994) 'An economic hothouse with amazing resilience', *Investors' Chronicle* 25 November: 75–6.

Griffiths, A. and Williams, A. M. (1992) 'Culture, regional image and economic development in the United Kingdom', *World Futures* 33: 105–29.

Griffiths, R. (1993) 'The politics of cultural policy in urban regeneration strategies', *Policy and Politics* 21, 1: 39–46.

—— (1995) 'Cultural strategies and new modes of urban intervention', *Cities* 12, 4: 253–65.

Gross, H. T. (1995) 'The role of health services in metropolitan and central city economic development: the example of Dallas', *Economic Development Quarterly* 9, 1: 80–6.

Guy, C. (1994) *The retail development process: location, property and planning*, London: Routledge.

Haggett, P., Cliff, A. D. and Frey, A. E. (1977) *Locational analysis in human geography*, London: Edward Arnold.

Hall, P. (1987) 'The anatomy of job creation: nations, regions and cities in the 1960s and 1970s', *Regional Studies* 21, 2: 95–106.

—— (1995) 'Towards a general urban theory', in J. Brotchie *et al.* (eds) *Cities in competition: productive and sustainable cities for the 21st century*, Melbourne: Longman.

—— (1996) 'The global city', *International Social Science Journal* 147: 15–24.

Hamilton-Fazey, I. (1993a) 'City strives to shine', *Financial Times* 9 June: 33.

—— (1993b) 'Grasping sharp nettles', *Financial Times* 9 June: 35.

Hamnett, C. (1994a) 'Social polarisation in global cities: theory and evidence', *Urban Studies* 31, 3: 401–24.

—— (1994b) 'Socio-economic change in London: professionalization not polarization', *Built Environment* 20, 3: 192–203.

Hansen, N. (1994) 'The strategic role of producer services in regional development', *International Regional Science Review* 16, 1–2: 187–95.

Harrington, J. W. (1995a) 'Empirical research on producer service growth and regional development', *The Professional Geographer* 47, 1: 66–9.

—— (1995b) 'Producer services research in U.S. regional studies', *The Professional Geographer* 47, 1: 87–96.

Harrington, J. W. and Lombard, J. R. (1989) 'Producer service firms in a declining manufacturing region', *Environment and Planning A* 21: 65–79.

Harrington, J., MacPherson, A. and Lombard, J. (1991) 'Interregional trade in producer services: review and synthesis', *Growth and Change* 22, 4: 75–94.

Harvey, D. (1989a) 'From managerialism to entrepreneurialism: the transformation in urban governance', *Geografiska Annaler* 71B, 1: 3–17.

—— (1989b) *The condition of post-modernity*, Oxford: Basil Blackwell.

Haug, P. (1995) 'Formation of biotechnology firms in the Greater Seattle region: an empirical investigation of entrepreneurial, financial and educational perspectives', *Environment and Planning A* 27: 249–67.

Haughton, G. (1996) 'Local leadership and economic regeneration in Leeds', in G. Haughton and C. C. Williams (eds) *Corporate City? Partnership, participation and partition in urban development in Leeds*, Aldershot: Avebury.

Haughton, G. and Hunter, C. (1994) *Sustainable cities*, London: Jessica Kingsley.

Haughton, G. and Williams, C. C. (1996) 'Corporate City?: A case of second city syndrome?', in G. Haughton and C. C. Williams (eds) *Corporate City?: Partnership, participation and partition in urban development in Leeds*, Aldershot: Avebury.

Heeley, J. (1996) 'Pulling in the people and the money', *Sheffield Star* 21 January: 7.

Heeley, J. and Pearlman, M. (1988) 'The Glasgow garden festival: making Glasgow miles better', *Fraser of Allander Institute Quarterly Economic Commentary* 14, 1: 65–70.

Hefner, F. L. (1990) 'Using economic models to measure the impact of sports on local economies', *Journal of Sport and Social Issues* 14: 1–13.

Henley Centre (1990) *Sport and the Welsh economy*, Cardiff: Sports Council for Wales.

—— (1992) *The economic impact of sport in the United Kingdom in the 1990s*, London: Sports Council.

Hepworth, M. E. (1989) *Geography of the information economy*, London: Belhaven.

Higher Education Funding Council for England (1995) *Higher education in further education colleges: funding the relationship*, Bristol: Higher Education Funding Council for England.

Hiles, D. R. (1992) 'Health services: the real jobs machine', *Monthly Labor Review* 115, 1: 3–16.

Hillier Parker (1991) *Shopping centres of Great Britain*, London: Hillier Parker.

Hohl, A. E. and Tisdell, C. A. (1995) 'Peripheral tourism: development and management', *Annals of Tourism Research* 22, 3: 518–34.

Home Office (1990) *The Hillsborough stadium disaster, 15th April 1989. Inquiry by Rt. Hon. Lord Justice Taylor*, London: HMSO.

House of Commons Employment Committee (1990) *Tourism: volume 1: report*, London: HMSO.

Howard, E. B. and Davies, R. L. (1987) *Retail change on Tyneside: third consumer survey in the MetroCentre*, OXIRM Research Paper no. A14, Oxford: Templeton College.

—— (1992) *Meadowhall: the impact of one year's development*, OXIRM Research Paper D9, Oxford: Templeton College.

Howard, E. B. and Reynolds, J. (1989) *Retail Development Fact Sheet*, Oxford: OXIRM, Templeton College.

Howells, J. (1988) *Economic, technological and locational trends in European services*, Aldershot: Avebury.

Hughes, A. (1996) 'Forging new cultures of food retailer-manufacturer relations' in N. Wrigley and M. Lowe (eds) *Retailing, consumption and capital: towards the new retail geography*, Harlow: Longman.

Hughes, C. G. (1982) 'The employment and economic effects of tourism reappraised', *Tourism Management* 3, 3: 167–76.

—— (1988) 'Conference tourism', *Tourism Management* 9, 3: 235–8.

Hughes, H. and Gratton, C. (1992) 'The economics of the culture industry', in D. Wynne (ed.) *The culture industry: the arts in urban regeneration*, Aldershot: Avebury.

Hughes, H. L. (1994) 'Tourism multiplier studies: a more judicious approach', *Tourism Management* 15, 6: 403–6.

Hunter, C. and Green, D. H. (1994) *Sustainable tourism*, London: Routledge.

Hunter, W. J. (1988) 'Economic impact studies: inaccurate, misleading and unnecessary', *Heartland Institute Policy Study 21*, Chicago: Heartland Institute.

Illeris, S. (1989a) *Services and regions in Europe*, Aldershot: Avebury.

—— (1989b) 'Producer services: the key sector for future economic development', *Entrepreneurship and Regional Development* 1: 267–74.

Illeris, S. and Sjoholt, P. (1995) 'The Nordic countries: high quality service in a low density environment', *Progress in Planning* 43, 2–3: 205–21.

Imbroscio, D. L. (1995) 'An alternative approach to urban economic development: exploring the dimensions and prospects of a self-reliance strategy', *Urban Affairs Review* 30, 6: 840–67.

International Labour Organization (1995) *World employment 1995*, Geneva: International Labour Organization.

Ioannides, D. (1995) 'Strengthening the ties between tourism and economic geography: a theoretical agenda', *The Professional Geographer* 47, 1: 49–60.

Jacobs, J. (1969) *The economy of cities*, New York: Random House.

Jefferson, A. (1995) 'Prospects for tourism: a practitioner's view', *Tourism Management* 16, 2: 101–5.

Johnson, A. T. (1986) 'Economic and policy implications of hosting sports franchises: lessons from Baltimore', *Urban Affairs Quarterly* 21, 3: 411–33.

—— (1989) *Local government and minor league baseball: survey of issues and trends*, Washington, DC: International City Management Association.

Johnson, P. and Thomas, B. (1990) 'Measuring the local employment impact of a tourist attraction: an empirical study', *Regional Studies* 24, 5: 395–403.

Johnson, R. L. and Moore, E. (1993) 'Tourism impact estimation', *Annals of Tourism Research* 20: 278–88.

Jones, B. D. and Vedlitz, A. (1988) 'Higher education policy and economic growth in the American States', *Economic Development Quarterly* 2, 1: 78–87.

Judd, D. R. and Ready, R. L. (1986) 'Entrepreneurial cities and the new politics of economic development', in G. E. Peterson and C. W. Lewis (eds) *Reagan and the cities*, Washington, DC: The Urban Institute Press.

Kaldor, N. (1966) *Causes of the slow rate of growth in the United Kingdom*, Cambridge: Cambridge University Press.

Keane, M. J. and Quinn, J. (1990) *Rural development and rural tourism*,

REFERENCES

Galway: Social Sciences Research Centre Research Report no. 5, University College Galway.

Kearns, G. (1993) 'The city as spectacle: Paris and the bicentenary of the French revolution', in G. Kearns and C. Philo (eds) *Selling places: the city as cultural capital, past and present*, London: Pergamon.

Keating, M. (1988) *The city that refused to die: Glasgow, the politics of urban regeneration*, Aberdeen: Aberdeen University Press.

—— (1991) *Comparative urban politics*, Aldershot: Edward Elgar.

Keeble, D., Tyler, P., Broom, G. and Lewis, J. (1992) *Business success in the countryside: the performance of rural enterprises*, London: HMSO.

Keil, S. and Mack, R. (1986) 'Identifying export potential in the service sector', *Growth and Change* 17: 1–10.

Kennedy, R. (1991) *London: world city moving into the 21st century*, London: HMSO.

Kidd, B. (1979) *The political economy of sport*, Ottawa: Canadian Association of Health, Physical Education and Recreation.

Kirklees Metropolitan Borough Council (1991) *Economic development strategy*, Huddersfield: Kirklees Metropolitan Borough Council.

Kotler, P., Haider, D. H. and Rein, I. (1993) *Marketing places*, New York: Free Press.

Lambert, J. (1996) 'Planning to take the city forward', *Sheffield Telegraph* 17 May: 43.

Landry, C. and Bianchini, F. (1995) 'The lifting of city limits', *The Guardian* 21 February: 20.

Lane, B. (1990) 'Will rural tourism succeed?', in S. Hardy, T. Hart and T. Shaw (eds) *The role of tourism in the urban and regional economy*, London: Regional Studies Association.

Lang, P. (1994) *LETS work: rebuilding the local economy*, Bristol: Grover Books.

Lash, S. and Urry, J. (1994) *Economies of signs and space*, London: Sage.

Law, C. (1990) 'Tourism as a focus for urban regeneration', in S. Hardy, T. Hart and T. Shaw (eds) *The role of tourism in the urban and regional economy*, London: Regional Studies Association.

—— (1992) 'Urban tourism and its contribution to economic regeneration', *Urban Studies* 24, 3–4: 599–618.

—— (1993) *Urban tourism: attracting visitors to large cities*, London: Mansell.

Lawless, P. (1986) 'Sheffield', in P. Cooke (ed.) *Global restructuring, local response*, Swindon: ESRC.

—— (1994) 'Partnership in urban regeneration in the UK: the Sheffield central area study', *Urban Studies* 31, 8: 1303–24.

—— (1995a) 'Inner city and suburban labour markets in a major English conurbation: processes and policy implications', *Urban Studies* 32, 7: 1097–1125.

—— (1995b) 'Spatial differentiation in urban employment patterns: inner city and suburban labour markets in an English provincial city', *Local Government Studies* 21, 2: 224–47.

Lawless, P. and Ramsden, P. (1990) *Sheffield into the 1990s: urban regeneration, the economic context*, Working Paper 23, Sheffield: Centre for Regional Economic and Social Research, Sheffield Hallam University.

268

Leatham, A. (1993) 'Going for gold', *Leisure Opportunities* 107: 32–3.

Leatherwood, T. (1983) 'Pension fund investment', in D. Jones and L. Webb (eds) *America's cities and counties: a citizen's agenda*, proceedings of conference on alternative state and local policies, Washington, DC.

Leeds City Council (1992) *Leeds: economic development strategy*, Leeds: Leeds City Council.

—— (1994) *Twenty-four hour city*, Leeds: Leeds City Council.

Leeds Development Agency (1994) *The Leeds economic development strategy: an economic assessment*, Leeds: Leeds Development Agency.

Leeds TEC Ltd (1995) *Leeds economic and labour market assessment 1995/96*, Leeds: Leeds TEC Ltd.

Leigh, C. (1995) *The Yorkshire and Humberside universities: a potential for the region*, Leeds: Yorkshire and Humberside Regional Research Observatory.

Leontidou, L. (1988) 'Greece: prospects and contradictions of tourism in the 1980s', in A. M. Williams and G. Shaw (eds) *Tourism and economic development: Western European experiences*, London: Belhaven.

Lever, J. (1973) 'Soccer in Brazil', in J. Talimini and C. Page (eds) *Sport and society: an anthology*, Boston, MA: Little, Brown and Co.

Lever, W. F. (1992) 'Local authority responses to economic change in West Central Scotland', *Urban Studies* 29, 6: 935–48.

Lewis, J. (1995) 'Seeing the light not blight', *Times Higher Education Supplement* 28 April: 13.

Ley, D. and Olds, K. (1988) 'Landscape as spectacle: world's fairs and the culture of mass consumption', *Environment and Planning C* 6: 191–212.

Leyshon, A. and Thrift, N. (1995) 'Geographies of financial exclusion: financial abandonment in Britain and the United States', *Transactions* 20, 3: 312–41.

Leyshon, A., Daniels, P. W. and Thrift, N. (1988) 'Large accountancy firms in the UK and spatial development', *Service Industries Journal* 8: 317–46.

Leyshon, A., Matless, D. and Revill, G. (1995) 'The place of music', *Transactions* 20, 4: 423–33.

Leyshon, A., Thrift, N. and Justice, M. (1993) *A reversal of fortune? Financial services and the south east of England*, Stevenage: South East Economic Development Strategy.

Light, D. and Prentice, R. (1994) 'Market-based product development in heritage tourism', *Tourism Management* 15, 1: 27–36.

Lim, H. (1993) 'Cultural strategies for revitalising the city: a review and evaluation', *Regional Studies* 27, 6: 589–95.

Lincoln, I., Stone, I. and Walker, A. (1994) 'The impact of higher education institutes on their local economy: a review of studies and assessment methods', Appendix II in J. Goddard *et al.*, *Universities and communities*, London: Committee of Vice-Chancellors and Principals.

Liversidge, D. (1995) 'Healthy regeneration', *Sheffield Telegraph: business review* 20 October: 7.

Loftman, J. (1991) *A tale of two cities: Birmingham the convention and unequal city – the International Convention Centre and disadvantaged groups*, Built

Environment Research Centre Research Paper no. 6, Birmingham: University of Central England.

London Tourist Board and Convention Bureau (1995) *London Tourist Board report and accounts 1994/95*, London: London Tourist Board and Convention Bureau.

—— (1996) *Recent trends in tourism*, London: London Tourist Board and Convention Bureau.

Lovatt, A. and O'Connor, J. (1995) 'Cities and the night-time economy', *Planning Practice and Research* 10, 2: 127–33.

Lowe, M. and Wrigley, N. (1996) 'Towards the new retail geography', in N. Wrigley and M. Lowe (eds) *Retailing, consumption and capital: towards the new retail geography*, Harlow: Longman.

Lowe, M. S. and Crewe, L. (1991) 'Lollipop jobs for pin money? Retail employment explored', *Area* 23: 344–7.

Lundmark, M. (1995) 'Computer services in Sweden: markets, labour qualifications and patterns of location', *Geografiska Annaler* 77B, 2: 125–39.

McCrone, G. (1991) 'Urban renewal: the Scottish experience', *Urban Studies* 28, 6: 919–38.

McGuigan, J. (1992) *Cultural populism*, London: Routledge.

MacInnes, J. (1995) 'The deindustrialization of Glasgow', *Scottish Affairs* 11: 73–95.

McKellar, S. (1988) 'The enterprise of culture', *Local Work* 8: 14–17.

McKenna, P. G. (1996) 'The research challenge faced by the new universities in the UK', *Higher Education Quarterly* 50, 2: 110–18.

McKillop, D. G., Ferguson, C. and Nesbitt, C. (1995) 'The competitive position of credit unions in the United Kingdom: a sectoral analysis', *Local Economy* 10: 48–64.

McLay, F. (ed.) (1988) *Workers' City: the real Glasgow stands up*, Glasgow: Clydeside Press.

—— (ed.) (1990) *Workers' City: the reckoning*, Glasgow: Clydeside Press.

McNulty, J. E. (1977) 'A test of the time dimension in economic base analysis', *Land Economics* 53: 358–68.

McWilliams, D. (1993) *London's contribution to the UK economy*, London: Corporation of London.

Maguire, J. (1995) 'Sport, national identities and globalization', in J. Bale (ed.) *Community, landscape and identity: horizons in a geography of sports*, Department of Geography Occasional Paper no. 20, Keele: University of Keele.

Mandelbaum, T. B. and Chicoine, D. L. (1986) 'The effect of timeframe in the estimation of employment multipliers', *Regional Science Perspectives* 12: 37–50.

Mansfield, E. (1991) 'Academic research and industrial innovation', *Research Policy* 20: 1–12.

Markie, P. (1990) 'The sunshine challenge', *Sheffield Telegraph* 21 September: 50.

Marsden, T. and Wrigley, N. (1996) 'Retailing, the food system and the regulatory state', in N. Wrigley and M. Lowe (eds) *Retailing, consumption and capital: towards the new retail geography*, Harlow: Longman.

REFERENCES

Marshall, A. (1919) *Industry and Trade*, London: Macmillan.

Marshall, J. N. (1979) 'Corporate organisation and regional office employment', *Environment and Planning A* 1: 553–63.

—— (1983) 'Business service activities in British provincial conurbations', *Environment and Planning A* 15: 1343–59.

—— (1996) 'Civil service reorganisation and urban and regional development in Britain', *Service Industries Journal* 16, 3: 347–66.

Marshall, J. N. et al. (1988) *Services and uneven development*, Oxford: Oxford University Press.

Marshall, J. N. and Raybould, S. (1993) 'New corporate structures and the evolving geography of white collar work', *Tijdschrift voor Economische en Sociale Geografie* 84: 362–76.

Marshall, J. N. and Richardson, R. (1996) 'The impact of "telemediated" services on corporate structures: the example of "branchless" retail banking in Britain', *Environment and Planning A* 28: 1843–58.

Marshall, J. N. and Wood, P. A. (1992) 'The role of services in urban and regional development: recent debates and new directions', *Environment and Planning A* 24: 1255–70.

—— (1995) *Services and space: key aspects of urban and regional development*, Harlow: Longman.

Marshall, J. N., Damesick, P. and Wood, P. A. (1987) 'Understanding the location and role of producer services in the United Kingdom', *Environment and Planning A* 24: 1255–70.

Marshall, J. N., Gentle, C. J. S., Raybould, S. and Coombes, M. (1992) 'Regulatory change, corporate restructuring and the spatial development of the British financial sector', *Regional Studies* 26: 453–68.

Massey, D., Quintas, P. and Wield, D. (1992) *High-tech fantasies: science parks in society and space*, London: Routledge.

Mathieson, A. and Wall, G. (1982) *Tourism: economic, physical and social impacts*, Harlow: Longman.

Matkin, G. W. (1994) 'Technology transfer and public policy: lessons from a case study', *Policy Studies International* 22, 2: 371–83.

Michalak, W. Z. and Fairbairn, K. J. (1993) 'The producer service complex of Edmonton: the role and organisation of producer services firms in a peripheral city', *Environment and Planning A* 25: 761–77.

Miller, H. and Jackson, R. (1988) 'The impact of the professional football strike on the Chicago land area', *Illinois Business Review* 45: 3–7.

Minihan, J. (1977) *The nationalisation of culture*, London: Hamish Hamilton.

Mohan, J. (1988a) 'Spatial aspects of health care employment in Britain: I, aggregate trends', *Environment and Planning A* 20: 7–23.

—— (1988b) 'Restructuring, privatisation and the geography of health care provision in England, 1983–87', *Transactions of the Institute of British Geographers* 13: 449–65.

—— (1991) 'The internationalisation and commercialisation of health care in Britain', *Environment and Planning A* 23: 853–67.

—— (1994) 'Universities as civic institutions: lessons from the USA', Appendix 3 in J. Goddard et al., *Universities and communities*, London: Committee of Vice-Chancellors and Principals.

Molotch, H. and Logan, J. R. (1985) 'Urban dependencies: new forms of use and exchange in US cities', *Urban Affairs Quarterly* 21, 2: 143–69.

Monopolies and Mergers Commission (1994) *The supply of recorded music: a report on the supply in the UK of pre-recorded compact discs, vinyl discs and tapes containing music*, London: HMSO.

Montgomery, J. (1990) 'Cities and the art of cultural planning', *Planning Practice and Research* 5, 3: 17–24.

—— (1995a) 'Urban vitality and the culture of cities', *Planning Practice and Research* 10, 2: 101–9.

—— (1995b) 'The story of Temple Bar: creating Dublin's cultural quarter', *Planning Practice and Research* 10, 2: 135–72.

Mooney, G. (1994a) 'Poverty in Glasgow', *Regions* 191: 11–14.

—— (1994b) 'The Glasgow regeneration alliance: the way forward for peripheral estates?', *Regions* 192: 13–16.

Morgan, K. (1982) 'Restructuring steel: the crisis of labour and locality in Britain', *International Journal of Urban and Regional Research* 17, 2: 175–201.

Moulaert, F. and Gallouj, G. (1995) 'Advanced producer services in the French space economy: decentralisation at the highest level', *Progress in Planning* 43, 2–3: 139–54.

Moulaert, F. and Todtling, F. (1995a) 'Preface', *Progress in Planning* 43, 2–3: 101–6.

—— (1995b) 'Conclusions and prospects', *Progress in Planning* 43, 2–3: 261–74.

Moulaert, F., Todtling, F. and Schamp, E. (1995) 'The role of transnational corporations', *Progress in Planning* 43, 2–3: 107–22.

Mulgan, G. and Warpole, K. (1992) *Saturday night or Sunday morning: from arts to industry-new forms of cultural policy*, London: Comedia.

Murphy, P. E. (1985) *Tourism: a community approach*, New York: Methuen.

Myerscough, J. (1988) *The economic importance of the arts in Britain*, London: Policy Studies Institute.

—— (1991) *Monitoring Glasgow 1990*, Report to Glasgow City Council, Strathclyde Regional Council and Scottish Enterprise.

National Heritage Select Committee (1995) *Bids to stage international sporting events: minutes of evidence Thursday 22 June 1995*, London: HMSO.

Newby, P. (1993) 'Shopping as leisure', in R. D. F Bromley and C. J. Thomas (eds) *Retail change: contemporary issues*, London: UCL Press.

Nilsson, P. A. (1993) 'Tourism in peripheral regions: a Swedish policy perspective', *Entrepreneurship and Regional Development* 5: 39–44.

North, D. (1955) 'Location theory and regional economic growth', *Journal of Political Economy* 63: 243–58.

Norton, E. (1993) 'Football at any cost: one city's mad chase for an NFL franchise', *Wall Street Journal* 13 October: 1.

Noyelle, T. (1983) 'The rise of advanced services', *American Planning Association Journal* 49: 280–90.

—— (1984) 'Rethinking public policy for the service era', *Economic Development Commentary* 8: 12–17.

O'Brien, K. (1990) *The UK tourism and leisure market*, London: The Economic Intelligence Unit.

O'Farrell, P. N. (1993) 'The performance of business service firms in peripheral regions: an international comparison between Scotland and Nova Scotia', *Environment and Planning A* 25: 1627–48.

—— (1995) 'Manufacturing demand for business services', *Cambridge Journal of Economics* 19, 4: 523–43.

O'Farrell, P. N. and Moffatt, L. A. R. (1995) 'Business services and their impact upon client performance: an exploratory inter-regional analysis', *Regional Studies* 29, 2: 111–24.

O'Farrell, P. N., Hitchens, D. M. and Moffatt, L. A. R. (1992) 'The competitiveness of business services and regional development: evidence from Scotland and the South East of England', *Urban Studies* 30: 1629–52.

—— (1995) 'Business service firms in two peripheral economies: Scotland and Ireland', *Tijdschrift voor Economische en Sociale Geografie* 86, 2: 115–28.

O'Farrell, P. N., Wood, P. A. and Zheng, J. (1996) 'Internationalization of business services: an inter-regional analysis', *Regional Studies* 30, 2: 101–18.

Offe, C. and Heinze, R. G. (1992) *Beyond employment: time, work and the informal economy*, Cambridge: Polity Press.

Office for National Statistics (1996) *New Earnings Survey 1995*, London: Office for National Statistics.

Okner, B. (1974) 'Subsidies of stadiums and arenas', in R. Noll (ed.) *Government and the sports business*, Washington, DC: Brookings Institute.

Organisation for Economic Co-operation and Development (1987) *Universities and local economic development*, Paris: OECD.

Pacione, M. (1993) 'The geography of the urban crisis: some evidence from Glasgow', *Scottish Geographical Magazine* 109, 2: 87–95.

—— (1995) *Glasgow: the socio-spatial development of the city*, Chichester: Wiley.

Paddison, R. (1993) 'City marketing, image reconstruction and urban regeneration', *Urban Studies* 30, 2: 339–50.

Page, S. (1990) 'Sport arena development in the UK: its role in urban regeneration in London docklands', *Sport Place* 4, 1: 3–15.

—— (1995) *Urban tourism*, London: Routledge.

Peck, J. and Tickell, A. (1991) 'Regulation theory and the geographies of flexible accumulation: transitions in capitalism; transitions in theory', *Spatial Policy Analysis Working Paper 12*, Manchester: School of Geography, University of Manchester.

Persky, J. and Wiewel, W. (1994) 'The growing localness of the global city', *Economic Geography* 70, 2: 129–43.

Persky, J., Ranney, D. and Wiewel, W. (1993) 'Import substitution and local economic development', *Economic Development Quarterly* 7: 18–29.

Peters, M. (1969) *International tourism: the economics and development of the international tourist trade*, London: Hutchinson.

Pieda (1991) *Sport and the economy in Scotland*, Edinburgh: Scottish Sports Council.

Poirier, R. A. (1995) 'Tourism and development in Tunisia', *Annals of Tourism Research* 22, 1: 157–71.

Polese, M. (1982) 'Regional demand for business services and inter-regional service flows in a small Canadian region', *Papers of the Regional Science Association* 50: 151–63.

Pollner, M. and Hollstein, J. (1985) 'Psychological well-being and the socio-symbolism of athletic contests', *Sociological Enquiry* 55: 291–309.

Port Authority of New York and New Jersey (1993) *The arts as an industry*, New York: Port Authority of New York and New Jersey.

Porterfield, S. and Pulver, G. C. (1988) *The export potential of service-producing industries: survey results*, Agricultural Economics Staff Paper no. 284, Washington, DC: Department of Agriculture Economic Research Service.

—— (1991) 'Service producers, exports, and the generation of economic growth', *International Regional Science Review* 14: 41–59.

Portsmouth City Council (1995) *D-Day and Le Tour: impact study and visitor survey*, Portsmouth: Portsmouth City Council.

Power, T. M. (1988) *The economic pursuit of quality*, Armonk, NY: Sharp.

Price, P. (1988) 'Have a little faith and vision', *Sheffield Star* 13 January: 9.

Public Policy Centre and SRI International (1986) *The higher education–economic development connection: emerging roles for public colleges and universities in a changing economy*, New York: American Association of State Colleges and Universities.

Quirk, J. (1973) 'An economic analysis of team movement in professional sport', *Law and Contemporary Problems* 38: 42–66.

Quirk, J. and Fort, R. D. (1992) *Pay dirt: the business of professional team sports*, Princeton, NJ: Princeton University Press.

Ragas, W. R., Miestchovich, I. J., Nebel, E. C., Ryan, T. P. and Lacho, K. J. (1987) 'Louisiana superdome: public costs and benefits 1975–84', *Economic Development Quarterly* 1, 3: 226–39.

Rakowski, C. A. (1994) 'Convergence and divergence in the informal sector debate: a focus on Latin America 1984–92', *World Development* 22, 4: 501–16.

Randall, J. E. and Warf, B. (1995) 'The impact of the AAG national meetings on community economies', paper presented to the Association of American Geographers Annual Meeting, Chicago, March.

Randall, J. N. (1980) 'Central Clydeside: a case study of one conurbation', in G. C. Cameron (ed.) *The future of British conurbations*, Harlow: Longman.

Redwood, A. (1988) 'Job creation in nonmetropolitan regions', *Journal of State Governments* 61: 9–15.

Regan, T. H. (1991) 'A study of the economic impact of the Denver Broncos football club on the Denver Colorado metropolitan economy', unpublished Ph.D. dissertation, University of Northern Colorado, Greeley.

Reynolds, J. and Howard, E. B. (1993) *The UK regional shopping centre: the challenge for public policy*, OXIRM Research Paper A28, Oxford: Templeton College.

Rice, T. (1995) 'Leeds leads the way', *Drapers' Record* 2 December: 19–22.

Richards, G. (1996) 'Production and consumption of European cultural tourism', *Annals of Tourism Research* 23, 2: 261–83.

Riddle, D. I. (1986) *Service-led growth: the role of the service sector in world development*, New York: Praeger.

Rifkin, J. and Barber, R. (1978) *The North will rise again: pension, politics and power in the 1980s*, Boston, MA: Beacon.

Ritchie, J. (1984) 'Assessing the impact of hallmark events: conceptual and research issues', *Journal of Travel Research* 23, 1: 2–11.

Robertson, J. (1989) *Future wealth: a new economics for the 21st century*, London: Cassell.

Robson, B., Deas, I., Topham, N. and Twomey, J. (1995) *The economic and social impact of Greater Manchester's universities*, Manchester: University of Manchester School of Geography.

Roche, M. (1992) 'Mega-events and micro-modernization: on the sociology of the new urban tourism', *British Journal of Sociology* 43: 563–600.

—— (1994) 'Mega-events and urban policy', *Annals of Tourism Research* 21: 1–19.

Rogerson, R., Findlay, A. and Morris, A. (1988) *A report on the quality of life in British cities*, Glasgow: University of Glasgow, Department of Geography.

Rolfe, H. (1992) *Arts festivals in the UK*, London: Policy Studies Institute.

Rosentraub, M. S. and Nunn, S. (1977) 'Suburban city investment in professional sport: estimating the financial returns of the Dallas cowboys and Texas rangers to investor communities', *American Behavioural Scientist* 21, 3: 393–414.

Rosentraub, M. S. and Swindell, D. (1991) 'Just say no? The economic and political realities of a small city's investment in minor league baseball', *Economic Development Quarterly* 5: 152–67.

Rosentraub, M. S., Swindell, D., Przybylski, M. and Mullins, D. R. (1994) 'Sport and downtown development strategy: if you build it, will the jobs come?', *Journal of Urban Affairs* 16, 3: 221–39.

Rouse, P. (1995) 'Sheffield's top 30 companies', *Sheffield Telegraph* 25 August: 1.

Rowley, G. (1993) 'Prospects for the central business district', in R. D. F. Bromley and C. J. Thomas (eds) *Retail change: contemporary issues*, London: UCL Press.

Sadek, J. (1995) 'Called to the bar', *Cities Management* March/April: 16–17.

Sassen, S. (1988) *The mobility of labour and capital*, Cambridge: Cambridge University Press.

—— (1991) *The global city: New York, London, Tokyo*, Princeton: Princeton University Press.

Sayer, A. and Walker, R. (1992) *The new social economy: reworking the division of labour*, Oxford: Blackwell.

Schaffer, W. A. and Davidson, L. S. (1984) *Economic impact of the Falcons on Atlanta*, Suwanee, GA: The Atlanta Falcons.

Schaffer, W. A., Jaffee, B. L. and Davidson, L. S. (1993) *Beyond the games: the economic impact of amateur sports*, Indianapolis: Chamber of Commerce.

Schamp, E. W. (1995) 'The geography of APS in a goods exporting economy: the case of West Germany', *Progress in Planning* 43, 2–3: 155–71.

Schuster, J. M. (1995) 'Two urban festivals: La Merce and First Night', *Planning Practice and Research* 10, 2: 173–87.

Scottish Tourist Board (1995a) *Overseas tourism in Greater Glasgow 1994*, Glasgow: Scottish Tourist Board.

Scottish Tourist Board (1995b) *British tourism in Greater Glasgow 1994*, Glasgow: Scottish Tourist Board.

Segal, Quince and Wicksteed (1985) *The Cambridge phenomenon*, Cambridge: Segal, Quince and Wicksteed.

—— (1988) *Universities, enterprise and local economic development: an exploration of the links*, London: HMSO.

Segebarth, K. (1990) 'Some aspects of international trade in services: an empirical approach', *Service Industries Journal* 10: 266–83.

Shaw, G. and Williams, A. M. (1994) *Critical issues in tourism: a geographical perspective*, Oxford: Blackwell.

Sheffield City Council (1987) *Sheffield's twin valleys: a strategy for economic regeneration*, Sheffield: DEED, Sheffield City Council.

—— (1995a) *Sheffield economic development plan 1995/6: proposed economic activities*, Sheffield: Sheffield City Council.

—— (1995b) *Sheffield economic development plan 1995/6: directory of major projects and initiatives*, Sheffield: Sheffield City Council.

Shields, R. (1991) *Places on the margin*, London: Routledge.

Short, J., Tremble, M., Fossu, K. and Child, D. (1990) *World Student Games: economic impact study, part I*, Sheffield: Sheffield City Council.

Smales, L. (1994) 'Desperate pragmatism or shrewd optimism? The image and selling of West Yorkshire', in G. Haughton and C. C. Williams (eds) *Corporate city? Partnership, participation and partition in urban development in Leeds*, Aldershot: Avebury.

Smales, L. and Whitney, D. (1996) 'Inventing a better place: urban design in Leeds in the post-war era', in G. Haughton and C. C. Williams (eds) *Corporate city? Partnership, participation and partition in urban development in Leeds*, Aldershot: Avebury.

Smales, S., Prior, J. and Burley, K. (1996) 'Sustainably flows the Don', *Planning Week* 4, 2: 16–17.

Smilor, R. W., Gibson, D. W. and Dietrich, G. B. (1990) 'University spin-out companies: technology start-ups from UT-Austin', *Journal of Business Venturing* 5: 63–76.

Smith, S. (1994) 'Groggy shopping', *The Guardian*, 22 December: 2.

Smith, S. L. J. (1989) *Tourism analysis: a handbook*, New York: Longman.

Smith, S. M. (1984) 'Export-orientation of non-manufacturing business in non-metropolitan communities', *American Journal of Agricultural Economics* 66: 145–54.

Smith, S. M. and Pulver, G. C. (1981) 'Non-manufacturing business as a growth alternative in nonmetropolitan areas', *Journal of the Community Development Society* 12, 1: 32–47.

Soane, J. V. N. (1993) *Fashionable resort regions: their evolution and transformation*, Wallingford: CAB International.

Squires, G. D. (1989) *Unequal partnerships: the political economy of urban redevelopment in postwar America*, New Brunswick, NJ: Rutgers University Press.

REFERENCES

Stabler, J. C. (1987) 'Nonmetropolitan population growth and the evolution of rural service centers in the Canadian prairie region', *Regional Studies* 21: 43–53.
Stabler, J. C. and Howe, E. C. (1988) 'Service exports and regional growth in the post-industrial era', *Journal of Regional Science* 28, 3: 303–16.
—— (1993) 'Services, trade and regional structural change in Canada, 1974–1984', *Review of Urban and Regional Development Studies* 5: 29–50.
Stigler, G. J. (1956) *Trends in employment in service industries*, New York: National Bureau for Economic Research, Princeton University Press.
Stillwell, J. and Leigh, C. (1996) 'Exploring the geographies of social polarisation in Leeds', in G. Haughton and C. C. Williams (eds) *Corporate city? Partnership, participation and partition in urban development in Leeds*, Aldershot: Avebury.
Stone, G. (1971) 'American sports: play and display', in E. Dunning (ed.) *The sociology of sport*, London: Cass.
Strange, I. (1993) 'Public–private partnership and the politics of economic regeneration policy in Sheffield, 1985–1991', unpublished Ph.D. thesis, University of Sheffield.
—— (1996a) 'Pragmatism, opportunity and entertainment: the arts, culture and urban economic regeneration in Leeds', in G. Haughton and C. C. Williams (eds) *Corporate city? Partnership, participation and partition in urban development in Leeds*, Aldershot: Avebury.
—— (1996b) 'Arts policy and economic regeneration in Leeds: pragmatism, opportunity and entertainment', *Local Government Policy Making* 22, 5: 30–8.
Swan, N. M. (1985) 'The service sector: engine of growth?', *Canadian Public Policy*, 344–50.
Targett, S. (1995) 'Campus boost to local wealth proven', *Times Higher Education Supplement* 28 April: 2.
Tarling, R., Rhodes, J., North, J. and Broom, G. (1990) *The economy and rural England*, London: Rural Development Commission.
Taylor, I., Evans, K. and Fraser, P. (1996) *A tale of two cities: global change, local feeling and everyday life in the North of England. A study in Manchester and Sheffield*, London: Routledge.
Taylor, N. (1990) 'Glasgow: a personal view of the city', *Catalyst* 3, 1: 2–4.
Taylor, P. (1989) *The economic impact of Bristol Polytechnic on local income and employment (1986/7)*, Bristol: Bristol Polytechnic, Department of Economics and Social Science.
Thomas, K. and Shutt, J. (1996) 'World Class Leeds: sectoral policy and manufacturing alliances in Leeds', in G. Haughton and C. C. Williams (eds) *Corporate city? Partnership, participation and partition in urban development in Leeds*, Aldershot: Avebury.
Thrift, N. J. (1990) 'Doing regional geography in a global system: the new international financial system, the City of London and the south east of England', in R. Johnson, J. Haver, and G. A. Hoekveld (eds) *Regional geography: current developments and future prospects*, London: Routledge.
Thrift, N. and Leyshon, A. (1992) *Making money: the City of London and social power in Britain*, London: Routledge.
Tickell, A. (1996) 'Taking the initiative: the Leeds financial centre', in G.

Haughton and C. C. Williams (eds) *Corporate city? Partnership, participation and partition in urban development in Leeds*, Aldershot: Avebury.

Tiebout, C. (1956) 'Exports and regional growth', *Journal of Political Economy* 64: 160–9.

Tight, M. (1996) 'The re-location of higher education', *Higher Education Quarterly* 50, 2: 119–37.

Todtling, F. (1984) 'Organisational characteristics of plants in core and peripheral regions of Austria', *Regional Studies* 18: 397–412.

Todtling, F. and Traxler, J. (1995) 'The changing location of advanced producer services in Austria', *Progress in Planning* 43, 2–3: 185–204.

Townsend, A. (1992) 'New directions in the growth of tourism employment? Propositions of the 1980s', *Environment and Planning A* 24: 821–32.

Townsend, A., Sadler, D. and Hudson, R. (1996) 'Geographical dimensions of UK retailing employment change', in N. Wrigley and M. Lowe (eds) *Retailing, consumption and capital: towards the new retail geography*, Harlow: Longman.

Tricker, M. and Williams, C. C. (1992) 'Breaking the rural mould', *Regional Studies Association Newsletter* 178: 14–22.

Turco, D. M. and Kelsey, C. W. (1992) *Conducting economic impact studies of recreation and parks special events*, Arlington, VA: National Recreation and Park Association.

Tym, R. and Partners (1993) *University of Greenwich: economic impact assessment*, London: Roger Tym and Partners.

United States Department of Agriculture (1993) *Rural conditions and trends*, Washington, DC: United States Department of Agriculture.

University of Sheffield (1995) *Annual report 1994/95*, Sheffield: University of Sheffield.

University of Southampton (1991) *Consequences of expansion of the University of Southampton*, Southampton: University of Southampton.

Urban Cultures Ltd (1992) *Manchester's cultural strategy*, Manchester: Manchester City Council.

—— (1994) *Prospects and planning requirements of the creative industries in London*, London: London Planning Advisory Group.

Urry, J. (1987) 'Some social and spatial aspects of services', *Environment and Planning D* 5, 1: 5–26.

—— (1990) *The tourist gaze: leisure and travel in contemporary societies*, London: Sage.

—— (1995) *Consuming places*, London: Routledge.

Van der Lee, P. and Williams, J. (1986) 'The Grand prix and tourism', in J. P. A. Burns, J. H. Hatch and T. J. Mules (eds) *The Adelaide Grand Prix: the impact of a special event*, Adelaide: Centre for South Australian Economic Studies.

Van Dinteren, J. J. (1987) 'The role of business service office in the economy of medium sized cities', *Environment and Planning A* 19: 669–86.

Vaughan, D. and Wilkes, K. (1986) *An economic impact study of tourist and associated arts development in Merseyside: the tourism study*, Liverpool: Merseyside Tourism Board.

REFERENCES

Vaughan, M., Hilsenrath, P. and Ludke, R. L. (1994) 'The contribution of hospitals to a local economy: a case study in Iowa and Illinois', *Health Care Management Review* 19, 3: 34–40.

Vaughan, R. (1990) 'Assessing the economic impact of tourism', in S. Hardy, T. Hart and T. Shaw (1990) *The role of tourism in the urban and regional economy*, London: Regional Studies Association.

Walker, M. (1995) 'Regional shopping centres: a critical evaluation of Princes Quay, Hull', unpublished MA dissertation in Town and Regional Planning, Leeds Metropolitan University.

Wainwright, M. (1995) 'Open all hours', *The Guardian* 7 June: 7.

Waple, A. (1996) 'The pain and the gain', *Sheffield Telegraph* 5 July: 18.

Warpole, K. (1992) *Towns for people: transforming urban life*, Buckingham: Open University Press.

Wavell, S. (1995) 'A tale of two cities', *Sunday Times* 11 June: 37.

Whitbread PLC (1994) *Cross-channel shopping: the facts*, London: Whitbread PLC.

Whitehead, A. (1993) 'Regional shopping centre development: a rejoinder', *Local Government Policy Making* 20, 2: 50–9.

Whitt, J. A. (1987) 'Mozart in the metropolis: the arts coalition and the urban growth machine', *Urban Affairs Quarterly* 23, 1: 15–36.

—— (1988) 'The role of the performing arts in urban competition and growth', in S. Cummings (ed.) *Business elites and urban development*, Albany: State University of New York Press.

Williams, A. and Shaw, G. (1990) 'Tourism in urban and regional development: Western European experiences', in S. Hardy, T. Hart and T. Shaw (1990) *The role of tourism in the urban and regional economy*, London: Regional Studies Association.

Williams, A., Shore, G. and Huber, M. (1995) 'The arts and economic development: regional and urban-rural contrasts in UK local authority policies for the arts', *Regional Studies* 29, 1: 73–80.

Williams, C. C. (1992a) 'The contribution of regional shopping centres to local economic development: threat or opportunity?', *Area* 24, 3: 283–8.

—— (1992b) 'Impacts of regional shopping centres: myths and realities', *The Planner* 78, 22: 8–11.

—— (1993) 'The implications of regional shopping centre development', *Local Government Policy Making* 20, 1: 50–5.

—— (1994a) 'Local sourcing initiatives in West Yorkshire: an evaluation of their effectiveness', in G. Haughton and D. Whitney (eds) *Reinventing a region: restructuring in West Yorkshire*, Aldershot: Avebury.

—— (1994b) 'Rethinking the role of the service sector in local economic revitalisation', *Local Economy* 9, 1: 73–82.

—— (1994c) 'Changing attitudes of government towards retail development: the end of the road out-of-town?', *Regional Review* 4, 2: 6–7.

—— (1994d) 'The role of the service sector in the economic revitalisation of West Yorkshire', in G. Haughton and D. Whitney (eds) *Reinventing a region: restructuring in West Yorkshire*, Aldershot: Avebury.

—— (1994e) 'A critical evaluation of the opposition to regional shopping centres', *Local Government Policy Making* 21, 3: 47–54.

REFERENCES

Williams, C. C. (1995) 'Opposition to regional shopping centres in Great Britain: a clash of cultures?', *Journal of Retailing and Consumer Services* 2, 4: 241–9.

—— (1996a) 'Understanding the role of consumer services in local economic development: some evidence from the Fens', *Environment and Planning A* 28, 3: 555–71.

—— (1996b) 'Consumer services and the competitive city', in G. Haughton and C. C. Williams (eds) *Corporate city? Partnership, participation and partition in urban development in Leeds*, Aldershot: Avebury.

—— (1996c) 'Rethinking the role of retailing and consumer services in local economic development: a British perspective', *Journal of Retailing and Consumer Services* 3, 1: 53–6.

—— (1996d) 'Local Exchange and Trading Systems (LETS): a new source of work and credit for the poor and unemployed?', *Environment and Planning A* 28, 8: 1395–1415.

—— (1996e) 'Local purchasing schemes and rural development: an evaluation of Local Exchange and Trading Systems (LETS)', *Journal of Rural Studies* 12, 3: 231–44.

—— (1996f) 'An appraisal of Local Exchange and Trading Systems (LETS) in the United Kingdom', *Local Economy* 11, 3: 275–82.

—— (1996g) 'Informal sector responses to unemployment: an evaluation of the potential of Local Exchange and Trading Systems (LETS)', *Work, Employment and Society* 10, 2: 341–59.

—— (forthcoming 1997) 'Rethinking the role of the retail sector in economic development', *Service Industries Journal* 17, 2.

Williams, C. C. and Thomas, R. (1996) 'Paid informal work in the Leeds hospitality industry: regulated or unregulated work?', in G. Haughton and C. C. Williams (eds) *Corporate city? Partnership, participation and partition in urban development in Leeds*, Aldershot: Avebury.

Williams, C. C. and Windebank, J. (1993) 'Social and spatial inequalities in informal economic activity: some evidence from the European Community', *Area* 25, 4: 358–64.

—— (1994) 'Spatial variations in the informal sector: a review of evidence from the European Union', *Regional Studies* 28, 8: 819–25.

—— (1995a) 'Social polarisation of households in contemporary Britain: a "whole economy" perspective', *Regional Studies* 29, 8: 727–32.

—— (1995b) 'Black market work in the European Community: peripheral work for peripheral localities?', *International Journal of Urban and Regional Research* 19, 1: 22–39.

—— (1996a) 'Shopping across frontiers: the economic implications of the growth in cross-channel shopping', in A. Church and P. Reid (eds) *Local economic frontiers: policy responses to 1992, the Channel Tunnel and the high speed rail links – Britain and France*, Aldershot: Avebury.

—— (1996b) 'The unemployed and informal sector in Europe's cities and regions', in P. Lawless, R. Martin and S. Hardy (eds) *Unemployment and social exclusion: landscapes of labour inequality*, London: Jessica Kingsley.

Williams, C. C., Turton, L., Howlett, S. and Thomas, K. (1993) *Economic impact assessment of the Earth Centre*, Earth Centre Working Paper no. 2, CUDEM, Leeds Metropolitan University.

280

REFERENCES

Wilson, A. G. (1974) *Urban and regional models in geography and planning,* Chichester: John Wiley.

World Tourist Organisation (1988) *Economic review of world tourism,* Madrid: World Tourist Organisation.

Wright, P. (1996) 'Threats to progress', *Sheffield Star* 29 January: 18.

Wrigley, N. (1994) 'After the store wars: towards a new era of competition in UK food retailing?', *Journal of Retailing and Consumer Services* 1, 1: 5–20.

Wrigley, N. and Lowe, M. (1996) (eds) *Retailing, consumption and capital: towards the new retail geography,* Harlow: Longman.

Wynne, D. (1992a) 'Cultural industries', in D. Wynne (ed.) *The culture industry: the arts in urban regeneration,* Aldershot: Avebury.

—— (1992b) 'Cultural quarters', in D. Wynne (ed.) *The culture industry: the arts in urban regeneration,* Aldershot: Avebury.

—— (1992c) 'Urban regeneration and the arts', in D. Wynne (ed.) *The culture industry: the arts in urban regeneration,* Aldershot: Avebury.

York, K. (1991) 'Sport and community identity: the case of Bath Rugby Football Club', *South Hampshire Geographer* 20: 12–23.

Yorkshire and Humberside Development Agency (1996) *Inward investment into Sheffield 1991–1996,* Leeds: Yorkshire and Humberside Development Agency.

Zimbalist, A. (1992) *Baseball and billions,* New York: Basic Books.

Zukin, S. (1982) *Loft living: culture and capital in urban change,* Baltimore: The Johns Hopkins University Press.

Zurick, D. N. (1992) 'Adventure travel and sustainable tourism in the peripheral economy of Nepal', *Annals of the Association of American Geographers* 82: 608–28.

INDEX

Mulgan, G. 132, 135
multiplier effects 64–6
Murphy, P.E. 59, 65
Myerscough, J. 6, 129–30, 141, 152, 158–9, 161–2, 230, 240

Netherlands 15, 16, 155
Newby, P. 119
Newcastle 187; *see also* MetroCentre
NIDL (new international division of labour) 18, 31
Nijkamp, P. 68
Nilsson, P.A. 204
non-services 222
Nordic countries 32: *see also* Denmark; Finland; Iceland; Norway; Sweden
North America: consumer services 50; cultural industries 128; deindustrialization and tertiarization 15; inter-regional and inter-local trade in services 21; medical services 181; producer services and economic regeneration 35; rural regeneration 210–11; sport and rejuvenation of local economies 74, 78; *see also* Canada; United States
North England 45, 76, 94, 110, 111, 121, 128, 137, 138, 139
North-West England 45, 76, 94, 110, 111, 128
Norton, E. 88
Norway 20, 30, 31, 33
Nottingham 187
Noyelle, T. 31
Nunn, S. 74

O'Brien, K. 58
O'Connor, J. 135
O'Farrell, P.N. 35, 36, 38
Offe, C. 244
Okner, B. 66, 80
Olds, K. 74
Organisation for Economic Cooperation and Development 19, 30, 59

Out-of-this-World initiative 124
Oxford 229, 237

Pacione, M. 148, 149–51
Paddison, R. 150–1, 152, 160
Page, S. 57, 78, 83, 89
Pearlman, M. 153–4
Peck, J. 13
Persky, J. 47, 49, 50, 213, 225
Peterborough 207
Peters, M. 61, 226
Pieda 75, 89
Poirier, R.A. 68
Polese, M. 25, 50, 202
Pollner, M. 80
Porterfield, S. 21, 204, 207, 210, 211, 217
Portugal 20, 30, 31–2, 220
Power, T.M. 49
Prentice, R. 57
Price, P. 172, 173
primary activities 2, 5, 41, 150
producer services and economic regeneration 3, 4–5, 6, 28–39, 150, 222: advanced producer services 28, 30, 31, 32, 33, 34, 39; export role 35–7; location 31–4; relationship to other sectors 37–9; role in economic development 34–9; size and growth 29–31
Producer Services Working Party 3
production 223
public services 3, 150, 222
Pulver, G.C. 21, 204, 207, 210, 211, 212, 217

Quesnais, F. 204
Quinn, J. 204
Quirk, J. 74, 84

Ragas, W.R. 84
Ramsden, P. 164, 165
Randall, J.E. 68–9, 71
Randall, J.N. 148
rationale for study of consumer services 5–8
Rawding, L. 57
Raybould, S. 220

of employees in employment
166; service sectors, fastest
growing 168; sports sector 83,
88, 170–3; steel, demise of 164–
7; tourism 68, 169–70; Travel-
to-Work-Area 165, 179;
universities 166, 167, 179–81;
Urban Programme and Regional
Policy 171; World Student
Games 168, 169, 170, 171–2,
173
Shields, R. 57
Short, J. 171
Shutt, J. 186
Singapore 18, 220
Single European Market 117–19,
120, 122, 123, 125
Sjoholt, P. 32–3
Smales, L. 140, 193, 199
Smilor, R.W. 100
Smith, G. 123
Smith, M.P. 220
Smith, S. 119
Smith, S.L.J. 79
Smith, S.M. 22, 204, 210, 211, 212,
217
Snee, H. 64
Soane, J.V.N. 58
South Africa 82
South England 138
South-East England 45, 76, 94,
110, 111, 121, 127, 128, 137
South-West England 45, 76, 94,
110, 111, 121, 128, 137
Spain 128, 134, 231
Spence, N. 219
sport and rejuvenation of local
economies 74–90: Britain 74–5,
77, 80–4, 88–90; contributions
to development 77–85; hallmark
events 81–3; magnitude of
sector 75–7; policy implications
88–90; regional distribution of
sports employment 76; regular
spectating events 78–81;
Sheffield 170–3; stadium
developments 83–5; United
States 78–81, 83, 84, 85–8
Squires, G.D. 85, 87

Stabler, J.C. 22, 24, 25, 48–9, 204,
207, 210, 211
standard industrial classification
index 4–5, 127, 151
Sternel, B.K. 22
Stigler, G.J. 2
Stillwell, J. 199
Stone, G. 74
Strange, I. 168, 195, 196
Sunderland 83
Swan, N.M. 26
Sweden 16, 19, 20, 30, 31
Swindell, D. 84
Switzerland 30

Targett, S. 107
Tarling, R. 202, 204
Taylor, I. 165, 177
Taylor, N. 152
Taylor, P. 98
Taylor Report 84
tertiarization 14–18
tertiary see services
Thailand 18, 220
Thomas, B. 64
Thomas, C. 177
Thomas, K. 186
Thomas, R. 57
Thrift, N. 113, 223, 243
Thurrock 113, 117
Tickell, A. 13, 185, 188, 191
Tight, M. 94
Tisdell, C.A. 65, 68, 72
Todtling, F. 19, 20, 28, 30, 31, 32,
33, 34, 35, 37, 38
tourism and economic
regeneration 55–73: Association
of American Geographers 68–
70, 71; Butlin's 66; Center Parcs
66; character 56–9; cultural and
sports 57; direct impacts 63–4;
enclave 65–6; geography in
Britain 60–2; Glasgow 160;
heritage 57; Hyatt-Regency 70;
implications for policy 71–3;
intangible impacts 66; local
economic impacts 62–70;
London 225–8; magnitude 59–
60; Marriott 70; multiplier

INDEX

West Midlands 45, 76, 94, 111,
    113, 117, 128
western Europe 15, 30
Whitbread PLC 118, 119
Whitehead, A. 114, 115
Whitney, D. 193, 199
Whitt, J.A. 126, 128, 134
Wiewel, W. 225
Wigan 139
Wilkes, K. 64
Williams, A. 59, 61, 62, 67, 137,
    138–9
Williams, A.M. 60, 131
Williams, C.C. 178–9, 187, 193–4,
    198, 205, 212, 234, 243–4: on
    consumer services 47, 49; on
    cultural industries 139; on
    manufacturing 22; on retailing
    108, 112, 113, 115, 116–17, 118,
    120, 124, 125; on tourism 55, 57;
    on universities 105

Williams, J. 79
Wilson, A.G. 5, 13
Windebank, J. 118, 120, 234
Wood, P.A. 2, 3, 225: on consumer
    services 41, 44, 45, 47, 49; on
    producer services 26, 33, 37, 38
Woolf, G. 219
Woolfson, C. 148
Wrigley, N. 110, 119
Wynne, D. 6, 126, 129, 135, 136,
    139

York 169, 198, 199, 200
Yorkshire and Humberside 45, 46,
    76, 94, 111, 128, 137, 245

Zimbalist, A. 84
Zukin, S. 139
Zurick, D.N. 67

293